Research at NaMLab

Band 3

Research at NaMLab

Band 3

Herausgeber:

Prof. Dr.-Ing. Thomas Mikolajick

Andreas Krause

Ultrathin Calcium Titanate Capacitors

Physics and Application

Logos Verlag Berlin

λογος

Research at NaMLab

Herausgegeben von
NaMLab gGmbH
Nöthnitzer Str. 64
D-01187 Dresden

Bibliografische Information der Deutschen Nationalbibliothek

Die Deutsche Nationalbibliothek verzeichnet diese Publikation in der
Deutschen Nationalbibliografie; detaillierte bibliografische Daten sind
im Internet über http://dnb.d-nb.de abrufbar.

ISBN 978-3-8325-3724-1
ISSN 2191-7167

Logos Verlag Berlin GmbH
Comeniushof, Gubener Str. 47,
10243 Berlin
Tel.:   +49 (0)30 / 42 85 10 90
Fax: +49 (0)30 / 42 85 10 92
http://www.logos-verlag.de

Technische Universität Dresden

Ultrathin CaTiO$_3$ Capacitors: Physics and Application

Andreas Krause
Dipl.-Phys.

von der Fakultät Elektrotechnik und Informationstechnik der Technischen Universität Dresden zur Erlangung des akademischen Grades eines

Doktoringenieurs
(Dr.-Ing.)

genehmigte Dissertation

Prüfungsvorsitzender: Prof. Dr. rer.nat. J.W. Bartha

1. Gutachter: Prof. Dr.-Ing. T. Mikolajick

2. Gutachter: Prof. Dr.-Ing. F. Kreupl

Einreichung: 01.11.2013

mündl. Prüfung: 23.01.2014

Verteidigung: 01.04.2014

Abstract

Scaling of electronic circuits from micro- to nanometer size determined the incredible development in computer technology in the last decades. In charge storage capacitors that are the largest components in dynamic random access memories (DRAM), dielectrics with higher permittivity (high-k) were needed to replace SiO_2. Therefore ZrO_2 has been introduced in the capacitor stack to allow sufficient capacitance in decreasing structure sizes. To improve the capacitance density per cell area, approaches with three dimensional structures were developed in device fabrication. To further enable scaling for future generations, significant efforts to replace ZrO_2 as high-k dielectric have been undertaken since the 1990s. In calculations, $CaTiO_3$ has been identified as a potential replacement to allow a significant capacitance improvement. This material exhibits a significantly higher permittivity and a sufficient band gap. The scope of this thesis is therefore the preparation and detailed physical and electrical characterization of ultrathin $CaTiO_3$ layers. The complete capacitor stacks including $CaTiO_3$ have been prepared under ultrahigh vacuum to minimize the influence of adsorbents or contaminants at the interfaces. Various electrodes are evaluated regarding temperature stability and chemical reactance to achieve crystalline $CaTiO_3$. An optimal electrode was found to be a stack consisting of Pt on TiN.

Physical experiments confirm the excellent band gap of 4.0-4.2 eV for ultrathin $CaTiO_3$ layers. Growth studies to achieve crystalline $CaTiO_3$ indicate a reduction of crystallization temperature from 640 °C on SiO_2 to 550°C on Pt. This reduction has been investigated in detail in transmission electron microscopy measurements, revealing a local and partial epitaxial growth of (111) $CaTiO_3$ on top of (111) Pt surfaces. This preferential growth is beneficial to the electrical performance with an increased relative permittivity of 55 with the advantage of a low leakage current comparable to that in amorphous $CaTiO_3$ layers. A detailed electrical analysis of capacitors with amorphous and crystalline $CaTiO_3$ reveals a relative permittivity of 30 for amorphous and an excellent value of 105 for fully crystalline $CaTiO_3$. The permittivity exhibits a quadratic dependence with applied electric field. Crystalline $CaTiO_3$ shows a 1-3% drop in capacitance density and permittivity at a bias voltage of 1 V, which is significantly lower compared to all results for $SrTiO_3$ capacitors measured elsewhere. A capacitance equivalent thickness (CET) below 1.0 nm with current densities 1×10^{-8} A/cm^2 have been achieved on carbon electrodes. Finally, CETs of about 0.5 nm with leakage currents of 1×10^{-7} A/cm^2 on top of Pt/TiN fulfill the 2016 DRAM requirements following the ITRS road map of 2012.

Zusammenfassung

Die Verkleinerung von elektronischen Bauelementen hin zu nanometerkleinen Strukturen beschreibt die unglaubliche Entwicklung der Computertechnologie in den letzten Jahrzehnten. In Ladungsspeicherkondensatoren, den größten Komponenten in Arbeitsspeichern, wurden dafür Dielektrika benötigt, die eine deutlich höhere Permittivität als SiO_2 besitzen. ZrO_2 wurde als geeignetes Dielektrikum eingeführt, um eine ausreichende Kapazität bei kleiner werdenen Strukturen sicherzustellen. Zur weiteren Verbesserung der Kapazitätsdichte pro Zellfläche konnten 3D Strukturen in die Chipherstellung integriert werden. Seit den 1990ern wurden parallel bedeutende Anstrengungen unternommen, um ZrO_2 als Dielektrikum durch Materialien mit noch höherer Permittivität zu ersetzen. Nach Berechnungen stellt nun $CaTiO_3$ eine mögliche Alternative dar, die eine weitere Verbesserung der Kapazität ermöglicht. Das Material besitzt eine deutlich höhere Permittivität und eine ausreichend große Bandlücke. Diese Arbeit beschäftigt sich deshalb mit Herstellung und detaillierter physikalischer und elektrischer Charakterisierung von extrem dünnen $CaTiO_3$ Schichten. Zusätzlich wurden diverse Elektroden bezüglich ihrer Temperaturstabilität und der chemischen Stabilität untersucht, um kristallines $CaTiO_3$ zu erhalten. Als eine optimale Elektrode stellte sich Pt auf TiN heraus. Physikalische Experimente an extrem dünnen $CaTiO_3$ Schichten bestätigen die Bandlücke von 4.0-4.2 eV. Wachstumsuntersuchungen an kristallinem $CaTiO_3$ zeigen eine Reduktion der Kristallisationstemperatur von 640 °C auf SiO_2 zu 550°C auf Pt. Diese Reduktion wurde detailliert mittels Transmissionselektronenmikroskopie untersucht. Es konnte für einige Schichten ein partielles lokales epitaktisches Wachstum von (111) $CaTiO_3$ auf (111) Pt gemessen werden. Dieses Vorzugswachstum ist vorteilhaft für die elektrischen Eigenschaften durch eine gesteigerte Permittivität von 55 bei gleichzeitig geringem Leckstrom vergleichbar zu amorphen Schichten. Eine genaue elektrische Analyse von Kondensatoren mit amorphen und kristallinem $CaTiO_3$ ergibt eine Permittivität von 30 für amorphe und bis zu 105 für kristalline $CaTiO_3$ Schichten. Die Permittivität zeigt eine quadratische Abhängigkeit von der angelegten Spannung. Kristallines $CaTiO_3$ zeigt einen 1-3% Abfall der Permittivität bei 1 V, der wesentlich geringer ausfällt als vergleichbare Werte für $SrTiO_3$. Eine zu SiO_2 vergleichbare Schichtdicke (CET) von unter 1.0 nm mit Stromdichten von 1×10^{-8} A/cm^2 wurde auf Kohlenstoffsubstraten erreicht. Mit Werten von 0.5 nm bei Leckstromdichten von 1×10^{-7} A/cm^2 auf Pt/TiN Elektroden erfüllen die $CaTiO_3$ Kondensatoren die Anforderungen der ITRS Strategiepläne für Arbeitsspeicher ab 2016.

Contents

Nomenclature

$\hbar$	Planck's constant $(h/2\pi)$
ε_0	Vacuum permittivity - $8.854188 \times 10^{-12} \frac{A\,s}{V\,m}$
e	Elementary charge - $1.602176 \times 10^{-19} C$
k	Relative permittivity, identical to κ or ε_r
k_B	Boltzmann constant - $1.380649 \times 10^{-23} J\ K^{-1}$
T_{an}	Annealing temperature
T_{crys}	Crystallization temperature
T_{dep}	Deposition temperature
AC	Alternating current
AFM	Atomic force microscopy
$CaTiO_3$	Calcium titanate
DRAM	Dynamic random access memory
EELS	Electron energy-loss spectroscopy
GIXRD	Grazing incidence X-Ray diffraction
HRTEM	High resolution TEM
HTXRD	High temperature XRD
IVT	Temperature dependent current measurement
LE-EELS	Low energy EELS
MIM	Metal-Insulator-Metal

PDA	Post-deposition anneal
PES	Photoemission spectroscopy
RBS	Rutherford Backscattering
RF	Radio frequency
RIE	Reactive Ion Etching
RMS	Rough mean square
RTP	Rapid thermal processing
SEM	Scanning electron microscopy
SMU	Standard Measuring Unit
SrTiO$_3$	Strontium titanate
TAT	Trap-assisted tunneling
TEM	Transmission electron microscopy
UHV	Ultra high vacuum - $10^{-7} - 10^{-10}$ mbar
UPS	Ultraviolet photoelectron spectroscopy
VUV	Vacuum ultraviolet spectroscopy
XFA	X-Ray fluorescence analysis
XPS	X-Ray photo emission spectroscopy
XRR	X-ray reflectance

1 Introduction

1.1 Motivation

With the ongoing increase in integration density of microelectronic circuits especially in dynamic random access memory (DRAM) circuits, conventional SiO_2 and even ZrO_2-based dielectrics have reached their limits for capacitor applications as their dielectric constants are not sufficiently high. To keep the capacitance C constant while shrinking the size of a circuit, the relative dielectric permittivity k has to be increased (Equation 1.1).

$$C = \varepsilon_0 \times k \, \frac{A}{d} \tag{1.1}$$

ε_0 is the vacuum permittivity, k the relative dielectric permittivity, A the area and d the thickness of the (plate) capacitor. Increasing the capacitance is possible with the

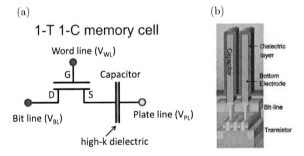

Figure 1.1: (a) Principle scheme of a 1-T 1-C memory cell in a DRAM containing a capacitor and a transistor. (b) Scheme of a memory cell as a 3D model. The capacitor is the largest part in the memory device (taken from Ref. [1]).

increase of the capacitance area and k and the reduction of the thickness d. The thickness scaling reaches the limit, when tunneling currents through the dielectric are the dominant conduction mechanism contributing to the leakage current. The capacitors are by far the

1

(a) 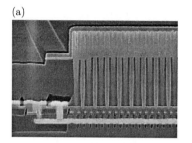   (b)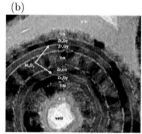

Figure 1.2: (a) A SEM cross-section image of a real device containing a row of capacitors and transistors. (b) Top TEM view of a high-k ZrO-AlO-ZrO (ZAZ) capacitor stack, as it is currently integrated in production. TiN is used as top and bottom electrode (taken from Ref. [2]).

largest components in DRAM memory cells (Figure 1.1 b)), therefore the capacitor size A prevents the further scaling of the devices.

Now, dielectric materials are required combining a high permittivity and low leakage currents for thin films. This is especially important for metal-insulator-metal (MIM) capacitors as they are the largest elements in DRAM cells. The principal arrangement of a 1-T 1-C memory cell stack is shown in Figures 1.1 a) and b). An example of a fully integrated DRAM cell with high-k dielectrics can be seen in Figure 1.2. Significant research has been done on alkaline earth metal compounds containing TiO_2 (or TiO_6 octahedrons) such as $SrTiO_3$ or $BaTiO_3$ as they can exhibit k values far beyond 100. However, these materials have not found their implementation in MIM capacitors due to processing and electrical performance issues. Both $SrTiO_3$ and $BaTiO_3$ exhibit a small band gap and $BaTiO_3$ also exhibits a ferroelectric phase transition temperature of 120°C. $CaTiO_3$ is a promising material for applications in MIM capacitors due to its relatively high permittivity ($k_{bulk} \sim 180$, orthorhombic) and a larger band gap (4.2 eV) compared to $SrTiO_3$ (3.6 eV) (Figure 1.3). In addition, the containing elements Ti an Ca both belong to the most abundant elements in Earth's crust.

As a figure of merit for capacitance of a MIM capacitor, the capacitance equivalent thickness (CET) is used to compare the high-k oxide layer to silicon dioxide according to Equation (1.2)

$$\frac{C}{A} = \varepsilon_0 \, k \, \frac{1}{d} = \varepsilon_0 \, k_{SiO2} \, \frac{1}{CET} \tag{1.2}$$

In this equation, A is the area of the capacitor, k_{SiO_2} the dielectric permittivity of SiO_2 ($k = 3.9$) and C is the capacitance of the high-k oxide layer.

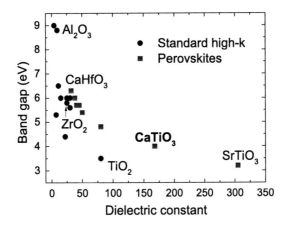

Figure 1.3: Bandgap vs. dielectric constant of various high-k oxides.

1.2 Structure of this work

Target of this work is the preparation and characterization of $CaTiO_3$ as an applicable dielectric material for capacitors in DRAM. Literature reveals no data on experiments of $CaTiO_3$ in capacitor arrangements for that purpose. The fundamental properties of $CaTiO_3$ are outlined in Section 2.1. To focus on the special requirements of $CaTiO_3$ given by the crystallization temperature above 600 °C and the band gap of 4.2 eV, the choice of electrodes are of major interest for a working capacitor. Noble metal electrodes exhibit various problems concerning preparation and temperature stability. The results of preparation of the various tested electrodes and diffusion barriers are presented in Chapter 4. The preparation of the full capacitor stacks including electrode and dielectric layer are described in detail in the experimental part of this thesis in Section 3.1. The physical properties of ultrathin $CaTiO_3$ layers have been investigated in detail in Chapter 5. The crystallization behavior of various $CaTiO_3$ capacitors has been investigated in detail with TEM in Chapter 6. The overall electrical performance of the fully prepared capacitors has been analyzed in Chapter 7. Both the capacitance behavior as well as the leakage currents have been studied in detail. The results prove the suitability of $CaTiO_3$ as an real alternative to $SrTiO_3$ in capacitor applications.

2 Fundamentals of materials and electrical properties

2.1 Calcium titanate

Materials in the perovskite structure ABO_3 are widely used in industry and research due to their intriguing properties like electroceramic dielectricity, superconductivity and ferroelectricity. With small changes to the crystal structure and composition with doping, the parameters of perovskites can be easily modified [3].

$CaTiO_3$ was the first material discovered in Russia in this ABO_3 composition [5]. Its

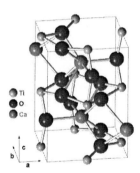

Figure 2.1: Unit cell of orthorhombic $CaTiO_3$ in Pbmn crystal structure at room temperature. The lattice parameters are a = 5.40 Å, b = 5.42 Å and c = 7.65 Å [4].

trivial name 'perovskite' therefore denotes the name of the oxide family with ABO_3 chemistry. With Ca^{2+} on the A side and Ti^{4+} on the B side, it belongs to the II-IV perovskite oxides group. The reason for its specific properties is described in the following sections.

2.1.1 Structure

The most simple structure for ABO_3 materials is a cubic phase. The cubic structure is typified by $SrTiO_3$ with a lattice constant of a=3.90 Å and the space group O_h^1-Pm3m. The Ti atoms are located at the corners and the Sr atoms at the center of the cube. The oxygen is placed at the centers of the twelve cube edges. The TiO_6 octahedra are perfect with 90° angles and six equal Ti-O bonds at 1.95 Å. Each Sr atom is surrounded by twelve equidistant oxygen atoms at 2.76 Å[3]. This structure is distorted to more complex structures depending on the chemical composition, temperature and pressure, e.g. in Ref. [6]. That distortion influences the dielectric properties as well as the band gap. Bulk $CaTiO_3$ crystallizes in an orthorhombic (Pbmn) unit cell at room temperature (Figure 2.1). This structure is energetically favorable at room temperature [7] in $CaTiO_3$ due to its tolerance factor[1] of $t \approx 0.95$. Phase transitions occur at temperatures above 1530 K towards a cubic Pm3m structure [8, 9]. The Pbnm structure is a result of condensation of two soft modes in Pm3m structure at R and M reciprocal lattice points. The TiO_6 octahedrons are tilted along the [001] and [010] direction [10]. The lattice constants of the $CaTiO_3$ orthorhombic Pbmn phase determined experimentally are a = 7.64 Å, b = 5.44 Å, c = 5.38 Å [5, 8] with its pseudocubic lattice constant of a = 3.82 Å.

2.1.2 Optical properties and band gap

As typical for all titanates ($ATiO_3$) including $CaTiO_3$, the band gap is formed by oxygen and titanium orbitals (see Figure 2.2). The formed TiO_6 octahedrons are tilted and distorted by the surrounding A atoms, resulting in different band gaps between $CaTiO_3$ and the other titanates. The band edges of the valence band arise from O 2s states. Hybridization of O 2p and Ti 3d orbitals lead to the conduction band edge [12, 13, 14]. The Ti−O transition establishes the band gap and is tuned by the surrounding grid of Ca atoms.

Various optical measurements were performed on thick $CaTiO_3$ layers to extract band gap and density of states (DOS) for occupied and unoccupied valence band and conduction band states respectively. Transmission measurements extracted a band gap of about 3.4 eV for bulk crystals [15]. For powders of $CaTiO_3$, a band gap has been measured between 3.86 eV and 3.97 eV [16, 17]. Thin layers ($\sim 100\,\mu m$) of $CaTiO_3$ show an optical band gap of 3.95 eV extracted from transmittance measurements [18].

---

[1]The tolerance factor is calculated from the ionic radii of the atoms A and B compared to the radius r_O of Oxygen: $t = \frac{r_A + r_O}{\sqrt{2}(r_B + r_O)}$. At a factor close to 1.0, cubic structure is preferred, while values $\neq 1$ result in a more distorted structure.

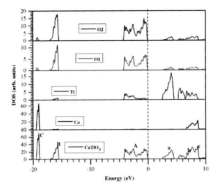

Figure 2.2: Calculated density of states for CaTiO₃ splitted by atoms (taken from [11]). The band gap of CaTiO₃ between valence band and conduction band is defined by the O2s orbitals forming the valence band edge and the Titanium 3d orbitals for the conduction band edge.

Density functional theory (DFT) calculations with various computing methods revealed band gaps near 2 eV for the Ti−O transitions, which is a typical underestimation for the used methods [9, 19]. Calculations with adapting semi-empirical fitting factors showed a band gap between 3.8-4.2 eV depending on the selected permittivity [9]. The largest optical band gap of 4.2 eV results have been recently calculated with a mixture of DFT and Hardtree-Fock methods in Ref. [20]. This method showed the best agreement to experimental values for other perovskites [21]. These values always lie above other literature values previously calculated for SrTiO₃ [12, 13, 15, 19].

2.1.3 Permittivity of CaTiO₃

CaTiO₃ exhibits a relatively high k-value of $k_{bulk} \approx 180$ in a bulk crystal [15, 23]. So far, no values for permittivity are found in literature for ultrathin layers of CaTiO₃. In simulations from Ref. [9] for supercells of up to 40 atoms, a k-value of 100 for CaTiO₃ has been calculated at room temperature, which is only exceeded by SrTiO₃.

The high permittivity or polarization of perovskites originates from dipole moments induced by vibrational modes of titanium against the oxygen ions leading to high dielectric constants. The large moments induced by these vibrational modes are linked to the geometrical arrangement of the Ti^{4+} ion within its oxygen octahedron in the CaTiO₃ unit cell. CaTiO₃, SrTiO₃ and BaTiO₃ show high-k constants, while in MgTiO₃ the Ti

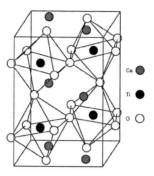

Figure 2.3: Distorted perovskite experimental room-temperature structure of $CaTiO_3$ (taken from Ref. [22]). The TiO_6 octahedrons are tilted along [100] and [010] directions.

ion only shares three oxygen ions [24] and therefore exhibits a k-value of only 27. The difference in dielectric permittivity of $CaTiO_3$ compared to $SrTiO_3$ is caused by the tilted TiO_6 octahedrons [10], which prevent any long-range coupling to neighboring unit cells (Figure 2.3).

With decreasing temperature, the $SrTiO_3$ cubic structure changes to more distorted unit cells. In principle, vibrational modes are frozen and permanent dipole moments arise in the crystal structure. Therefore the dielectric constant increases with decreasing temperature. The temperature where a phase transition occurs is called the Curie - Weiss temperature. Materials above the Curie - Weiss temperature, which show a decreasing permittivity with increasing temperature, are named incipient ferroelectrics [25]. This effect is shown for $CaTiO_3$ in Ref. [5] as well as for other materials of the same family, e.g. $SrTiO_3$, $KTaO_3$ and also for TiO_2 in Refs. [26, 27]. Typical dielectrics exhibit a decrease of the dielectric constant with decreasing temperature. For incipient ferroelectrics, the temperature dependence of the permittivity ε can be described according to Barret et al. in Ref. [28] with

$$\varepsilon(T) = \frac{C}{\frac{T_1}{2} \coth\left(\frac{T_1}{2T}\right) - T_0} \quad .$$

(2.1)

Here, T_1 is the temperature of the crossover between classical and quantum behavior, T_0 is the Curie - Weiss temperature of phase transition and C is the Curie - Weiss constant. Without impurities, $CaTiO_3$ has a calculated phase transition temperature $T_0 <$ $0\,K$ [29]. In $SrTiO_3$, only quantum fluctuations associated to a non-negligible zero point energy prevent the onset of ferroelectric long-range order [30]. With the introduction of impurities, $SrTiO_3$ exhibits a ferroelectric transition above $40\,K$. No ferroelectric transition has been observed for $CaTiO_3$ doped with concentrations below $30\ \%$ with Ba or Pb down to temperatures of $4\,K$. That means, the paraelectric phase is more stable in $CaTiO_3$ than in $SrTiO_3$. Structural calculations of T_0 reveal a Curie-Weiss temperature for phase transition to lie at $T_0 = -111\,K$ for $CaTiO_3$ [5] and $40\,K$ for $SrTiO_3$. Phase transitions occur with high amounts of impurities in $CaTiO_3$, e.g. Pb in $CaTiO_3$ [29] and Sr in $CaTiO_3$ [31], which induce a distinct change in atom-atom distances within the $CaTiO_3$ unit cell.

Recent experimental research investigates stress induced ferroelectricity in $CaTiO_3$. Stress-induced long-range order of dipole moments at twin walls showed an antiferrielectric phase of $CaTiO_3$ [32]. Ferroelectricity has also been observed on $CaTiO_3$ surfaces [33]. Current transmission electron microscopy experiments show ferrielectric displacements of Ti atoms at a (110) twin boundary of $CaTiO_3$ [34]. Also, recent simulations are investigating the stress-induced ferrolelectricity of orthorhombic $CaTiO_3$ [35, 36].

2.1.4 Current applications for $CaTiO_3$

$CaTiO_3$ is currently used in a variety of applications in industry and research, using its properties described above and taking advantage of the dielectric properties and the good lattice match to e.g. $SrTiO_3$ and other perovskites.

- **Transparent semiconductor**
  Doping $CaTiO_3$ with 3^+ or 5^+ ions changes the conductivity from an insulator to a semiconductor or show even metallic behavior. $CaTiO_3$ is selected due to the good lattice match to the widely used $SrTiO_3$ and $LaAlO_3$ substrates. The doping with Nb as donor material introduces a trap state near the conduction band [18, 37, 38].

- **Microwave applications**
  $CaTiO_3$ is due to its relatively high Q value (reciprocal of dielectric loss tangent $\tan\rho$) and sufficiently high permittivity at microwave frequencies (2-20 GHz) widely used in its pure state or as dopant in dielectric resonators or filters. It is often mixed with $MgTiO_3$ to further reduce the dielectric loss. Therefore $CaTiO_3$-based ceramics have wide applications as resonators, filters and antennas for communication, radar and global positioning systems operating at microwave frequencies [39, 40, 41, 42].

- **Laser application**
  Pr doped $CaTiO_3$ can be excited with light in the UV region with a wavelength

in the range from 250 to 500 nm. The unique emission line is at 614 nm. This material is favorable due to its weak afterglow and high chemical stability at room temperature [37, 38, 43].

- **Bioactive layers with CaTiO₃**
  Perovskites have been studied extensively for biomechanical applications due to their low Young's modulus and high elastic limit. To prevent toxic elements like Ni, Al and Be in biological applications, $CaTiO_3$ is of great interest in this area. $CaTiO_3$ increases the adherence of hydroxyapatite (basis for bones or teeth) to Ti prostheses [44]. Typically, the $CaTiO_3$ layers for this applications are prepared using sol-gel processes with a typical thickness in the micrometer range [45, 46].

The presented applications for $CaTiO_3$ describe the current knowledge of research. So far, no literature of $CaTiO_3$ in nanoelectronic devices has been found. The research of ultrathin $CaTiO_3$ layers for integration in thin film capacitors is therefore the topic of this thesis. The results of $CaTiO_3$ allow the comparison with $SrTiO_3$ and $BaTiO_3$ layers with their different properties to complete the picture of the ABO_3 materials group.

2.2 Metals and conducting metal composites for applicable high temperature processes

In high-k dielectrics, the band gap between valence band and conduction band is inverse proportional to the k-value (Figure 1.3)[47]. The small band gap results in a low band offset between the Fermi level of the metal and the conduction band of the dielectric. At room temperature, a considerable amount of electrons have sufficient thermal energy to overcome the barrier (Schottky emission). This is an important mechanism of leakage currents through the capacitor, that limits charge retention. The effect is even enhanced when a bias voltage is applied and the effective barrier is lowered. The effects of barrier lowering and emission into the conduction band is described in detail in Section 2.4.1.

To increase the barrier height between high-k dielectrics and electrodes, metals or conducting metal composites with high work functions are needed. At room temperature, a barrier height above 1 eV is commonly considered to be sufficient to block most carrier diffusion/injection. An excerpt of possible materials is given in Table 2.1. To be mentioned, high work function metals have to fulfill additional physical requirements. Parameters like temperature stability, reactivity towards oxygen, structuring suitability are important to produce metal-insulator-metal (MIM) stacks. As industrial requirement, the material costs have to be taken into account as well.

- **Platinum** is due to its high work function of 5.6 eV very suitable as an electrode material. Its lattice constant of 3.92 Å is comparable to a variety of high-k materials like $SrTiO_3$ and $CaTiO_3$ and can work as seeding layer without an interfacial layer or lattice adaption. In contradiction, the materials high cost prevents its use in industrial processes. Pt also reacts easily with Si and forms silicides ($PtSi_x$) at temperatures below 300 °C. Additionally, the high chemical etch resistance of Pt prevents a simple structuring process.

- **Ruthenium** is a noble metal of the Platinum group and possesses a work function of 4.6 eV [48]. The material costs are not as high as for Pt, which makes Ru an alternative and has been considered for application in industry processes. The pure metal reacts easily with oxygen starting at 280 °C and forms Ruthenium-(IV)-oxide (see below). The lattice expansion hinders Ru for use in MIM capacitors with device temperatures above 280 °C. This reactivity towards oxygen allows the use as a structured top electrode. Ru can be etched with reactive ion etching (RIE) of oxygen plasma under the formation of volatile Ruthenium tetraoxide (RuO_4)(see sample preparation in section 3.1.4). A wet etching process for Ru is available with (ortho-)periodic acid ($HIO_4 + 2H_2O \rightleftharpoons H_5IO_6$)[49].

- **Ruthenium-(IV)-oxide** (RuO_2) is an conductive oxide with a higher work function compared to Ru of about 5.0-5.2 eV [48, 50, 51]. As an oxide, it is more resistant to further oxidation and therefore shows increased stability at elevated temperatures. It can be structured using the same process mentioned above by the formation of volatile RuO_4. Other conducting materials containing Ru are **Strontium/ Calcium ruthenate** ($SrRuO_3$ and $CaRuO_3$), which crystallize in a distorted perovskite structure and therefore allows the oriented crystal growth of various perovskites like $SrTiO_3$. The materials are stable at high temperatures, show considerable conductivity and crystallize at 450 °C [48, 52, 53, 54].

- **Titanium nitride** (TiN) is a conductive ceramic material with a high work function of 4.4-4.7 eV and high temperature stability. It has been widely adapted to industrial processes as electrode material in various electronic devices or as protection layers in common mechanical tools. It is uniformly deposited by CVD, PVD and ALD. TiN works as a diffusion barrier for Si and metal atoms. Nevertheless, at elevated temperatures, it easily oxidizes to form TiON, an insulator with a low k-value. Wet etching processes are available to structure TiN electrodes [55].

Table 2.1: Summary of properties of suitable electrode materials according to Ref. [56].

Material	Work function (eV)	Lattice constant (Å)	Lattice structure	Etching characteristics	Oxidation resistance
Pt	5.6	3.92	fcc	difficult	yes
Ru	4.6	2.70	hexagonal	simple	no
TiN	4.6	1.6	cubic	simple	no
RuO$_2$	5.2	6.38 [110]	tetragonal	simple	yes
		3.11 [001]			
		4.49 [010]			

2.3 Non-linearity of dielectric permittivity

In various dielectrics like perovskites or other complex material compositions, physical properties including the permittivity are strongly anisotropic in different crystal orientations. The polarization of the dielectric is therefore dependent on the local electric field and the local polarizability. With mesoscopic capacitors used for characterization these attributes become homogeneously distributed throughout the capacitor. For a perfect, homogeneous and isotropic dielectric, the Clausius-Mosotti equation (Equation (2.2)) was the first connection between the local polarizability and the macroscopic permittivity.

$$\frac{k-1}{k+2}\frac{M_m}{\rho} = \frac{N_A}{3\varepsilon_0}\,\alpha_P \qquad (2.2)$$

Here, k is the dielectric permittivity, M_m is the molar mass, ρ is the density, N_A is the Avogadro's number, ε_0 is the vacuum permittivity and α_P as the polarizability of the dielectric. This uniformity can be applied for amorphous layers, where the material shows isotropy in permittivity in all directions. In polycrystalline dielectric layers, the properties are an intermixture of amorphous and (single-)crystalline properties. With the existence of preferential crystal directions the properties may shift towards single crystalline layers. When these crystallites are randomly distributed, the layers may show an isotropic behavior comparable to an amorphous material.

The polarization $\boldsymbol{P}$ of a dielectric material is described to be

$$\mathbf{P} = \varepsilon_0\chi\mathbf{E} \qquad (2.3)$$

with ε_0 as the electric permittivity of free space, χ as the electric susceptibility and $\mathbf{E}$ as the electric field. Only in a homogenous, linear and isotropic dielectric, the polarization is proportional to the electric field, otherwise χ describes a three-dimensional tensor. Capacitance voltage $(C(V))$ measurements typically integrate over a large capacitor area, therefore the integral over local inhomogeneities result in a homogenous dielectric and can be ruled out. Isotropy of the material can be expected for an amorphous dielectric and shall be excluded for crystalline dielectric layers.

An applied external electric field can result in an induced nonlinear polarization. Materials exhibiting this behavior are called nonlinear dielectrics (e.g. $CaTiO_3$, $SrTiO_3$ or $BaTiO_3$). A constant susceptibility will result in a linear dependence of polarization from an applied field. The relationship between polarization $\mathbf{P}$ and capacitance $C(V)$ (divided by the area A) is given in Equation (2.4).

$$\mathbf{P} = \frac{1}{A}\int C(V)dV \qquad (2.4)$$

In the case of a constant susceptibility, the capacitance has to be voltage/electric field independent. Literature for other perovskites shows non-linearity in $C(V)$, therefore comparable signatures are expected for CaTiO$_3$ layers. Typical $C(V)$ curves can be estimated with low error using a quadratic approximation (Equation (2.5)).

$$C(V) = C_0 \left(1 + \beta V + \alpha V^2 \right) \qquad (2.5)$$

with C_0 as the extremal capacitance, β as the linear voltage coefficient of capacitance (VCC) of C_0 and α as the quadratic VCC in dependence of the applied electric field. Various theoretical approaches as explanations have been tested for different materials, three approaches for comparable systems to CaTiO$_3$ are described in more detail.

At first, the quadratic VCC in $C(V)$ measurements has been attributed to the electrostriction [57]. The model only works for amorphous dielectrics with isotropic electrostriction, while for crystallites the anisotropic electrostrictive coefficient prevents further comparison. The α-factor for electrostriction as the main non-linear component is than described by

$$\alpha = \frac{2n_2 n_0}{kd^2} \qquad (2.6)$$

where α is dependent on the refractive index n_0, the linear dielectric constant k, the nonlinear refractive index n_2 and the thickness of the capacitor d [57]. This model has been validated for amorphous high-k dielectrics like Al$_2$O$_3$, HfO$_2$ or Y$_2$O$_3$. No frequency or temperature dependence is described within this model.

Second, different theories explain the dependence of capacitance from voltage with ionic or orientation polarization models. In Refs. [58, 59], a polarization model describes the variation of dielectric constant with the local metallic cation displacement in an externally applied electric field. This model has been tested on Al$_2$O$_3$ and shows a linear increase of capacitance with increasing temperature as well as a linear increase of the quadratic VCC with capacitance. In contradiction, the negative quadratic VCC of SiO$_2$ [60, 61] has been calculated with an orientation polarization model that shows a reduction of polarization with applied electric field (at 100 kHz) [62]. The magnitude of the negative quadratic VCC was found to be inversely proportional to the square of the sample's thickness. An equation based on the orientation polarization of the dipole moments has been derived from these results.

Approaches to combine both negative and positive dependencies of quadratic VCC with applied electric field have been investigated in Refs. [63, 64]. The introduction of nonlinear potentials revealed a short-range interaction with nearest neighbors for amorphous SrTiO$_3$ and long-range potentials for crystalline samples, resulting in positive quadratic VCC and negative quadratic VCC respectively. Typically, amorphous layers show a reduced atom-atom distance with a reduced number of nearest neighbors, while for crystalline samples, the atom-atom distance increases with more nearest neighbors. This changes the local field of each atom and is the reason for differences between amorphous and crystalline dielectric layers.

Another reasonable model uses the electrode polarization with the formation of a double-layer capacitance at the electrode interface [65, 66, 67, 68]. Here, the oxygen affinity of electrodes is important to control the magnitude of non-linearities. Various high-k dielectrics ($BaTiO_3$, HfO_2) and electrodes have been tested, showing the increase of (positive) non-linearity of capacitance with increase of oxygen affinity (free energy of metal oxidation per oxygen atom or heat of formation).

2.4 Leakage current mechanisms

In metal-insulator-metal (MIM) capacitors, various mechanisms are possible for leakage currents through the insulator and for discharging the capacitor. A comprehensive schematic description of these mechanisms is given in Figure 2.4. One can differentiate between conduction with or without defects/traps in the bulk insulator. Without defects, Schottky emission (SE) and tunneling of charge carriers are responsible for leakage currents. When defects are introduced, additional mechanisms like Poole-Frenkel (PF) emission and trap-to-trap tunneling (TAT) play a major role in charge carrier conduction through the dielectric. The complete leakage current in high-k MIM capacitors is than the sum of these mechanisms (Equation (2.7)).

$$I_{high-k} = \underbrace{I_{SE} + I_{Tunneling}}_{\text{w/o defects}} + \underbrace{I_{PF} + I_{TAT}}_{\text{with defects}} \qquad (2.7)$$

These conduction models are described briefly in the following text.

2.4.1 Schottky emission

As shown in Figure 1.3, the band gap of a dielectric is generally inverse proportional to the dielectric permittivity. High-k materials exhibit a rather low band gap compared to SiO_2 and Al_2O_3, therefore Schottky emission is evident when using low work function electrodes. The inclusion of surface trap states further induce the Schottky emission as one of the major leakage current mechanisms. In an ideal metal-insulator interface, the

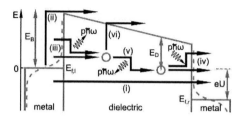

Figure 2.4: Schematic band diagram of a MIM structure with applied voltage U. Possible charge transport mechanisms: (i) direct/Fowler - Nordheim tunneling, (ii) thermionic emission, (iii) elastic/inelastic tunneling into and (iv) out of traps, (v) trap-to-trap-tunneling, (vi) PF emission (taken from Ref. [69]).

barrier height Φ_B is simply the difference between the metal work function Φ_M and the electron affinity χ_A of the insulator.

$$\Phi_B = \Phi_M - \chi_A \tag{2.8}$$

This ideal case is never found experimentally. The barrier height is changed with an additional interface layer and interface states as well as image forces at the interface [70]. When traps in the dielectric are persistent (by defects or doping), a trap-band is formed which reduces the effective barrier height as well.

The image force is described as force induced from an electron in a specific distance x from the Fermi level of the metal surface. An attractive force is introduced comparable to a mirror charge at a distance $(-x)$. With an external field E, the potential energy E_{pot} is given by

$$E_{pot}(x) = -\frac{1}{4\pi\varepsilon_0\varepsilon_i}\frac{q^2}{4x} - q\,|\mathbf{E}|\,x \quad . \tag{2.9}$$

q is the elementary charge and ε_i the effective permittivity in the dielectric. The position of the maximum value at x_M and image force lowering $\Delta\phi$ is therefore

$$x_M = \sqrt{\frac{q}{16\pi\varepsilon_0\varepsilon_i\,|\mathbf{E}|}} \tag{2.10}$$

$$\Delta\phi = \sqrt{\frac{q\,|\mathbf{E}|}{4\pi\varepsilon_0\varepsilon_i}} = 2\,|\mathbf{E}|\,x_M \tag{2.11}$$

The Schottky barrier is considerably lowered for low k materials. ε_i is expected to be significantly larger in high-k dielectrics and the barrier lowering could be neglected. Nevertheless, if the electron transit time to the barrier maximum is shorter than the dielectric relaxation time, the insulator is not polarized and ε_i is smaller than the permittivity k in the low frequency range. A typical carrier velocity is in the order of 10^7 cm/s. The transit time for the distance to the barrier height is $<10^{-14}$ s. The image force dielectric constant ε_i should than be comparable to ε_∞ for electric radiation in that range (optical constant for visible light)[70].

The current I_{SE} of Schottky emission is now given by

$$I_{SE} = A \ A^{**} T^2 \exp\left(\frac{q\left(\Phi_B - \sqrt{q\mathbf{E}_i/4\pi\varepsilon_i}\right)}{k_B T} \right) \tag{2.12}$$

with A as the area of the capacitor, A^{**} as the effective Richardson constant[2], Φ_B as barrier height, $\mathbf{E}_i$ as the electric field in insulator and ε_i as the effective dielectric permittivity (usually equal to ε_∞).

2.4.2 Tunneling

With a finitely high potential barrier and a finite thickness d, electrons possess a nonzero probability to tunnel through that barrier. This was the main reason of leakage currents for devices containing pure SiO$_2$ with thicknesses below 4 nm [71]. For thicker oxide layers, tunneling can be neglected due to the exponential decay of tunneling probability. The voltage/field dependence of the tunneling current is given by Equation 2.13 [70]

$$I_{Tunneling} \propto \exp\left(-\frac{4}{3} \frac{\sqrt{2qm^*}}{\hbar} \frac{\Phi_B^{3/2}}{\mathbf{E}} \right) \quad . \tag{2.13}$$

Here, q is the elemental charge, m^* is the effective mass, Φ_B is the barrier height and $\mathbf{E}$ is the electric field.

In a capacitor stack, the effective tunneling barrier thickness d can be reduced with a large electric field, typically above 10^9 V/m, forming a triangular barrier. This tunneling of electrons from the Fermi level of the metal to the conduction band of the oxide is called Fowler-Nordheim-Tunneling.

2.4.3 Poole-Frenkel effect

In high-k dielectrics, there are high amounts of bulk defects (oxygen vacancies or structural defects) in the order of 10^{18} cm^{-3} [69] or above. This makes the defect sensitive

---

[2]The effective Richardson constant for a free electron $A^{**} = \frac{4\pi q m^* k_B^2}{h^3}$ and $m^* = m_0$ is 120 A cm^{-2}K^{-2}.

leakage mechanisms play an important role in leakage currents for various high-k capacitors, when the energy level of the defect lies within the band gap of the dielectric.
The Poole-Frenkel effect (field-assisted thermal emission) is the lowering of a coulombic potential barrier when it interacts with an electric field. This is usually associated with the lowering of a trap barrier in the bulk of an insulator [72]. At sufficiently high trap densities, conduction through these traps is responsible for leakage currents at higher temperatures. **Poole-Frenkel emission** occurs, when a trapped electron within the band gap of the oxide layer is emitted into the conduction band. The behavior is comparable to Schottky emission where the barrier height Φ_B now describes the trap depth. The current density for Poole-Frenkel emission I_{PF} is described as

$$I_{PF} \propto \mathbf{E}_i \exp \frac{q\left(\phi_B - \sqrt{q\phi_i/\pi\varepsilon_I}\right)}{k_B T} \quad . \tag{2.14}$$

The Poole-Frenkel emission does not describe the probability of an electron tunneling into a specific trap. Statistical models comparable to **Trap-assisted Tunneling** (TAT) are needed to calculate the saturation current [69, 73].

2.4.4 Trap-assisted tunneling

Trap-assisted tunneling (TAT) is typically the main conduction mechanism, when high defect densities are present in the oxide layer. This is valid for materials with huge oxygen deficiencies or at grain boundaries between crystallites with different orientation. TAT strongly depends on the defect density and the defect distribution as well as the defect energy dept. The energy levels of oxygen in dependence on the vacancy charge, extracted from measurements of HfO_2 capacitors, lie between 0.8-1.2 eV below the conduction band [74, 75].

At low temperatures, temperature induced conduction mechanisms can be neglected. Than, TAT via deep traps is the dominant leakage current mechanism [76, 77]. For higher temperatures, electron hopping occurs into traps below the conduction band [77] and is strongly temperature dependent. In the case of inter-trap tunneling, the TAT current can be described as Equation (2.15) according to Refs. [78, 79]

$$I_{TAT} \propto \frac{1}{w^2} \exp\left(-\frac{2\sqrt{2m^*q}\sqrt{\phi_t}\,w}{\hbar}\right) \sinh\left(\frac{\sqrt{2m^*q}\,w^2\mathbf{E}}{\hbar\sqrt{\phi_t}}\right) \quad . \tag{2.15}$$

w is the distance between the nearest traps and ϕ_t the energy position of the traps. As conclusion, TAT is important in materials with high defect densities. It contributes mainly at low applied fields to the leakage current and shows negligible temperature dependence.

As a note, the different leakage current models mentioned earlier have been developed for dielectric materials with thicknesses above 50 nm, low k below 10 and low trap densities. When going to ultrathin high-k dielectrics with a smaller band gap and high trap densities, these models may not necessarily adaptable to dielectrics recently investigated.

2.5 Dielectric relaxation

Dielectric relaxation effects are observed in all dielectric materials including perovskites [80]. The effect is defined as "the decay of polarization from an initial steady state to zero after sudden removal of an initial polarizing field" [81]. The effect is besides leakage currents responsible for a charge loss in capacitors [82]. After a capacitor is charged, relaxation current leads to a voltage drop at the capacitor. The empirical Curie−von Schweidler law follows the Equation (2.16).

$$I(t) \propto \left(\frac{t}{t_0}\right)^{-n} \quad (n \approx 1; \;\; 0 < n < 2) \; . \tag{2.16}$$

The dielectric relaxation can be modeled with a series of RC elements[3] parallel to the basic RC element [83]. The Equation (2.16) results in a linear decrease in a log-log plot of current density vs. time. With increasing charging time, the relaxation currents are overshadowed by the static leakage currents. The physical reason of additional RC elements is not clear, one explanation is a time dependent tunneling current in a double-well potential [81].

2.6 Amorphous versus crystalline dielectric layers

The conduction and valence bands arise from overlaps of atomic orbitals, where the band structure is mainly dependent on the local structural order. This difference is only small between crystalline and amorphous solids. Small differences in amorphous materials arise from bond length and angle of bonds, which result in blurring of band gaps. The structural disorder also result in defect states in the band gap, which may result in changes of leakage current (Hunklinger 2007 p.404)[84]. Models of leakage currents are mainly developed for simple crystalline or polycrystalline samples. These systems give an excess for extensive theoretical simulations. Leakage currents in these layers are attributed to

---

[3]Electric circuit composed of a resistor and a capacitor

conduction along grain boundaries [85, 86, 87].

The situation is different for amorphous layers. While there are a lot of structure related defects, the layers are typically much smoother than crystalline layers. The presence of defects and impurities leads to Frenkel-Poole type leakage currents [88, 89]. TAT for amorphous layers with very high trap densities have also been observed [90]. In amorphous layers, vacancies and defects are homogeneously distributed through the complete oxide and not concentrated at grain boundaries. Therefore no leakage paths along grain boundaries contribute to the overall leakage current.

3 Sample Preparation and Methods

3.1 Sample Preparation of metal-insulator-metal capacitors

3.1.1 Sputtering of electrodes and high-k dielectrics

In this work, a customized ultra high vacuum (UHV) three chamber sputter cluster (Figure 3.1) is used to deposit the metals and oxide materials. The machine is capable to process up to 150 mm wafers. Typically, five quadratic 16 cm^2 substrates are simultaneously prepared using an adapter plate. All samples are baked out under vacuum for eight hours at 150 °C in the load lock chamber prior to transfer in the deposition chambers to ensure an UHV environment. The various layers are uniformly deposited onto the employed substrates. A schematic view of MIM capacitor preparation is shown in Figure 3.2. Extensive literature can be found regarding the physics of sputtering and influence of process parameters like pressure and power. A well written book for the basics of sputtering is Ref. [91].

3.1.2 Standard parameters for metal deposition

By default, metal layers are deposited between 100 °C and 300 °C in the metal chamber. The distance target surface - substrate center is approx. 180 mm. This large distance for sputter deposition ensures a metal layer deposition with ±10 % deviation from the mean thickness. Rotation of the substrate is essential for an uniform deposition of the selected material. The rotation is set to 6 rounds per second (rpm), which is sufficient for the applied deposition rate. The pressure range for deposition in this work is set between 1.2-1.6×10^{-3} mbar, which is at the lower range of typical sputtering pressures. Higher deposition rates and higher particle energy are chosen to achieve the best results for the materials system. Further sputtering parameters are summarized in Table 3.1.

Oxide chamber 1   Oxide chamber 2

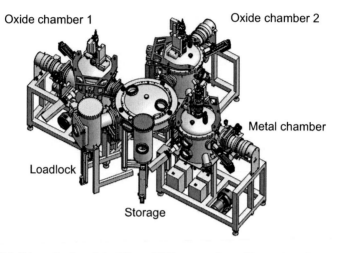

Metal chamber

Loadlock

Storage

Figure 3.1: Schematic view of the 150 mm UHV sputter cluster (Bestec GmbH) used to prepare the samples. It contains two oxide chambers and one metal chamber. Ten
samples can be loaded at once into the load lock. In addition, ten samples can be
stored in the storage chamber. An automatic handling system is used to transport
the substrates without breaking the vacuum to the different chambers (Ref. [92]).

3.1.3 Parameters for dielectric deposition

The dielectric deposition is done in a comparable chamber like the metal deposition
chamber (see previous part). With radio frequency (RF) magnetron sputtering technique,
$CaTiO_3$ and other oxides are sputtered from stoichiometric oxide targets. The deposition
temperature has been varied between room temperature and 700 °C. The distance target
surface - substrate center has been set to approx. 240 mm as well as a rotation speed
of 6 rpm to achieve layer uniformities below ±5 % deviation. The deposition rate with a
sputtering power of 120 W is about 0.25 nm/min at the selected distance. The pressure
range for deposition lies at 1.2×10^{-3} mbar. No oxygen as reactive gas has been introduced
during deposition because of the strong decrease of deposition rate. This leads to an
oxygen deficiency in the as-deposited layers. Nevertheless, experiments with additional
oxygen during sputtering showed no significant change of the already very low leakage
currents.

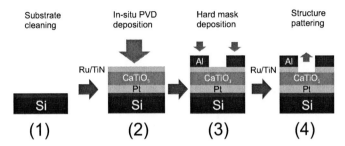

Figure 3.2: From substrate to structured MIM capacitors. (1) Clean substrates are baked out under UHV for 8 h at 150 °C. (2) The deposition of electrodes and CaTiO₃ is done *in-situ* without breaking vacuum. (3) For structuring an Al/Ti or Pt/Ti hard mask is deposited ex-situ. (4) The uncovered top electrode is removed to get a fully structured MIM capacitor.

3.1.4 Structuring of top electrodes

For electrical characterization of the prepared capacitors, a structured top electrode is required. The structuring has been realized by ex-situ processes. These are carried out with the deposition of a hard mask (circles with different diameters between 100 μm and 500 μm, Figure 3.2(3)) on top of the top electrode layer using thermal evaporation of different metals.

With Ru as a noble metal top electrode, the structuring has been performed using Reactive Ion Etching (RIE)[1] with an Ar-O plasma mixture. An Ti (200 nm) and Al (approx. 300 nm) stack has proven to be a suitable hard mask for RIE. The formation of a thin and hard Al_2O_3 cover layer prevents the further oxidation of Al and Ti. The uncovered Ru

---

[1]RIE of Ru was done at the Institute of Semiconductors and Microelectronics (IHM) by C. Richter.

Table 3.1: Table of deposition parameters for metal layer deposition.

Material	Plasma power (W)	Sputtering pressure ($\times 10^{-3}$ mbar)	Gas flow (sccm)	Deposition temperature (°C)
Pt	100	1.6	22	300
Ru	100	1.2	22	100
TiN	200	1.6	$\frac{20(Ar)}{4.5(N_2)}$	300
RuO$_2$	200	1.2	$\frac{20(Ar)}{4(O_2)}$	150

is etched with RIE under the formation of volatile Ruthenium-(IV)-oxide (RuO_4)(Figure 3.2(4)). This structuring is selective towards $CaTiO_3$. With no end point detection revealing the start of the $CaTiO_3$ layer, the minimum oxide thickness with this method has been determined to a thickness of 15 nm to achieve capacitors with reliable leakage currents. Dry etching of Pt as an alternative top electrode is more elaborate and not selective towards $CaTiO_3$. A more controlled etching process would be needed instead. With TiN as top electrode, standard wet etching processes are available to structure the MIM capacitor. An alkaline etching solution consisting of 500 ml deionized water, 20 ml Hydrogen peroxide (H_2O_2) and 10 ml Ammonia solution is heated to 50 °C. The hard mask stack used for RIE of Ru is not applicable, therefore Ti (200 nm) and Pt (10 nm) have been employed as a new hard mask. The samples with 15 nm TiN are etched in 5 min down to the oxide layer. Because the etching rate of TiN is significantly larger compared to the rate for $CaTiO_3$, the etching can easily be stopped without affecting the oxide layer. The controlled etching process allowed a structuring of capacitors with oxide thicknesses down to 8 nm with low leakage currents.

3.2 Physical characterization

Various tools have been utilized to characterize the physical properties of the $CaTiO_3$ MIM capacitor stacks. Important properties needed are the physical thickness of the samples as well as the degree of crystallinity of the $CaTiO_3$ layer. A standard **UV-vis spectroscopic ellipsometer** (Sopra) has been used in the range between 250 nm and 800 nm wavelength to extract the physical thickness. The ellipsometry setup also allows to determine the basic ellipsometric parameters Δ and Ψ at a fixed angle of incidence (usually 75°). To extract the optical constants for different layer thicknesses and crystallization states, the incident angle has been varied from 75° to 65°. To do so, an optical n,k-file for the sputtered $CaTiO_3$ has been created (see appendix). With an opaque top electrode in complete capacitor stack, calibration samples of pure $CaTiO_3$ layers on Si have been prepared to evaluate the thickness.

X-ray measurements are done with a Bruker Discover 8 using Cu k_α radiation and a Göbel mirror for parallel beam path. To characterize the thickness, the samples are evaluated in **X-ray reflectance** geometry (XRR). Samples with simple layer structure (in maximum 2 layers on Si substrate) have been utilized to extract reliable physical properties. **Grazing Incidence X-ray Diffraction** (GIXRD) is used to extract the crystallinity of the capacitor stack, in particular from the oxide. Temperature dependent XRD (high temperature XRD - HTXRD) measurements reveal an *in-situ* change of crystallinity of the measured samples. The measurements were applied by Dr. L. Wilde at Fraunhofer CNT. A Beryllium hemisphere has been used, which encapsulates the samples in a pure nitrogen atmosphere during heating.

The roughness and layer morphology of all sputtered layers is analyzed using **atomic force microscopy** (AFM) (Veeco Dimension 3100) as well as **scanning electron microscopy** (SEM) (Zeiss LEO 1560). The rough mean square (RMS) extracted from AFM images of $CaTiO_3$ as well as the electrode layers is used as an estimation of layer roughness. Smooth electrode layers are preferred for capacitors applications to prevent field emission of electrons on spikes due to high electric fields. An RMS below approx. 1.0 nm is considered to be suitable for electrical characterization. With both AFM and SEM, crystallinity of the layers, grains as well as cracks of the layers due to density increase after annealing can be observed.

Photoemission spectroscopy (PES) as a well-established method in surface and interface physics is used to study the electronic (valence band electronic states in the vicinity of the chemical potential, injection barriers, work function) and chemical (chemical composition, environment and reactions, oxidation states) properties of atoms, molecules, solids as well as their surfaces and interfaces. PES is an experimental technique probing the occupied states of a given system. In this kind of experiment an electron is excited into vacuum with a certain kinetic energy that depends mainly on its binding energy. It is possible to perform core level spectroscopy when a core level electron is excited by X-rays (XPS - X-ray photoelectron spectroscopy) and valence band spectroscopy by exciting valence band electrons with UV-light (UPS - ultraviolet photoelectron spectroscopy), respectively. PES allows the determination of electron binding energies and electron density at the sample surface. As a very surface sensitive technique PES has an information depth of 1 - 2 nm. Extensive literature regarding PES can be found in the monographs of D. Briggs and J. T. Grant [93], G. Ertl *et at.* [94], S. Hüfner [95], M. Cardona and L. Ley [96] and H. Ibach [97], respectively. The XPS measurements have been done externally at the Leibniz Institute for Solid State and Materials Research Dresden (IFW) by Dr. M. Grobosch.

For the combined X-ray photoemission (XPS) and ultraviolet photoemission spectroscopy (UPS), a commercial PHI 5600 spectrometer has been used. Photons with an energy of 1486.6 eV from a monochromatized Al K_α source for XPS and photons from a helium discharge lamp with an energy of 21.21 eV for UPS measurements are provided. All ultraviolet photoemission spectroscopy measurements were done by applying a sample bias of -9 V to obtain the correct, sample determined, secondary electron cutoff. The recorded spectra were corrected for the contributions of He satellite radiation. The total energy resolution of the spectrometer was determined by analyzing the width of the Au Fermi edge to be about 350 meV (XPS) and 100 meV (UPS), respectively. The binding energy (BE) scale was aligned by measuring the Fermi edge (0 eV) and the Au $4f_{7/2}$ emission feature (84.0 eV) of a polycrystalline, atomically clean gold substrate.

Transmission electron microscopy (TEM)[2] has been used to investigate the local physical properties like crystallinity and band gap of the complete MIM capacitors. Using High Resolution TEM (HRTEM), it was possible to extract the orientation of $CaTiO_3$ towards its substrate. In addition, with an included **low energy electron energy-loss spectroscopy** (LE-EELS) setup, the electronic band gap of the $CaTiO_3$ has been investigated [98, 99].

Extensive literature regarding physical characterization methods can be found in Refs. [56, 100, 101, 102].

3.3 Electrical characterization

For electrical characterization of the deposited $CaTiO_3$ layers, measurements are performed on a MIM capacitor stack to investigate the electrical properties. An in-house probe station (Süss PA200) for up to 200 mm diameter wafers is used for the 16 cm^2 structured samples. A parameter analyzer (Keithley instruments Model 4200 Semiconductor Characterization System) is connected directly to the bottom electrode via the probe station chuck and to the top electrode with a measurement tip.

The connection of the bottom electrode via the substrate requires a conducting substrate for reliable MIM capacitor characterization. As conducting substrate, wafers with either a high dopant concentration or a gas phase As doped Si surface are available. To further increase the connection quality, all samples are backside covered with an Al layer (200-300 nm thickness).

3.3.1 Current voltage measurements

Measurements of current versus voltage characteristics of high-k capacitors is more complex than capacitors containing SiO_2 or Al_2O_3 as dielectric materials. The high amount

---

[2]HRTEM was done using a Titan^3 FM 80-300 at the Leibniz Institute for Solid State and Materials Research Dresden (IFW) by Dr. D. Pohl. Further TEM to determine crystallinity has been done at the Dept. of Applied Physics at Eindhoven Univ. of Technology by Dr. M.A. Verheijen.

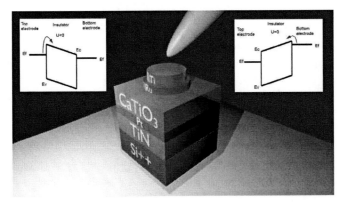

Figure 3.3: Principle measurement scheme of the patterned capacitors using a TiN/Pt bottom
electrode. (inset left) When negative voltage is applied to the top electrode,
electrons are injected through the top electrode. (inset right) For positive voltages
on the top electrode, the electrons are injected through the bottom electrode. This
is due to different electrodes the most interesting setup for current versus voltage
characterization.

of traps or defects in a high-k capacitor stack show a high temperature dependence of
leakage currents, which have to be taken carefully into account. With their inhomogeneous degree of crystallinity including grain boundaries, the applied external electric field
is therefore not homogeneously distributed over the whole capacitor. Different degree of
crystallinity in the CaTiO$_3$ layer can apply in a parallel manner (equivalent to capacitors
with different capacitance in a parallel arrangement) or even more complex structure,
when crystallites are inhomogeneous perpendicular to the applied potential (equivalent
to capacitors in serial arrangement). These difficulties in field distribution are comple-
mented with the complex behavior of grain boundaries. Due to stress and high defect
density [103], they can be responsible for leakage paths through the layer [86, 87] or show
changes in electrical behavior (e.g. ferrielectric twin walls in CaTiO$_3$ grain boundaries
[32]). These local effects are responsible for difficulties extracting single leakage mechanisms or break down voltages for feature sizes of 100 μm and above. Therefore, every
measurement is always the sum of various effects affecting the leakage current.

For measurement of leakage current of the prepared MIM capacitors, the measuring unit
(SMU) is connected to both the substrate and the patterned top electrode. In this arrangement, positive voltage applied to the top electrode is equivalent to the injection of

electrons through the bottom electrode. Negative voltage applied means the injection of electrons through the top electrode. This principle measurement scheme can be seen in (Figure 3.3). As delay time between each applied voltage step, five seconds are chosen. This is a compromise between measurement time and the influence of relaxation currents mainly contributing at low voltages. In insulators, the hole mobility in the valence band is much lower compared to electrons in the conduction band and hole conduction is typically neglected [104]. Furthermore due Fermi level pinning, the metals Fermi level is closer to the conduction band and therefore, the band offset is smaller for electrons [105].

3.3.2 Leakage current measurements at different temperatures

For standard characterization, leakage current measurements are typically done at room temperature (25°C). With the usage of temperature dependent IV measurements (IVT) between -40 °C and 200 °C, it is possible to observe changes in conduction mechanisms and extract relevant physical properties. Arrhenius plots are useful to visualize temperature dependent leakage currents. For that, the coordinates are translated according to $1/k_B T$ as X and $\ln j$ as Y axis to linearize the temperature dependence. Any dependence in the Arrhenius equation (Equation (3.1)) is the result of an exponential dependence on temperature.

$$j = A \exp^{\frac{-E_a}{k_b T}} \tag{3.1}$$

with j as leakage current density, A as a pre-exponential factor and E_a as an arbitrary activation energy. As main interest, E_a shall be extracted from IVT with the transformation of Equation (3.1) to

$$\ln \frac{j}{A} = \ln \frac{A}{A} - \frac{E_a}{k_B T} \qquad . \tag{3.2}$$

The activation energy is now the gradient of IVT in the Arrhenius plot mentioned above. Any pre-exponential factor A causes a constant vertical shift in Arrhenius plots.

3.3.3 Capacitance voltage measurements

Standard capacitance voltage $(C(V))$ measurements are done on the semiautomatic probe station used for I(V) measurements. The used Keithley 4200-CVU multi-frequency impedance measurement card is directly connected to this probe station. The high potential is connected to the top electrode, the low potential is connected to the substrate. A measurement takes place automatically on the $16\,cm^2$ structured samples. The large number of capacitors with different top electrode diameters allow the measurement of unused capacitors (typically $I(V)$ and $C(V)$ measurements are done on every second capacitor). As standard setting, a frequency of 100 kHz and a low measurement speed for low-noise results is used ('Quiet' mode; includes an internal SMU filter and automatic integration time setting, delay time is set to 0 ms).

The CVU measures impedance by sourcing an AC voltage across the sample, and then measures the resulting AC current and phase difference. The capacitance and conductance are derived parameters from the measured impedance and phase [106]. The capacitance C of the capacitor is calculated out of the capacitive impedance and the test frequency according to Equation (3.3)

$$C = \frac{I_c}{2\pi f V_c} \tag{3.3}$$

with I_c as the current, f as measured frequency and V_c as the AC amplitude [106]. The impedance Z is given as

$$Z = X + iY \tag{3.4}$$

with X as the resistance and Y as the reactance. The measured phase angle Φ and impedance allows the calculation of both X and Y with Equations (3.4) and (3.5).

$$\Phi = \arctan \frac{X}{Y} \tag{3.5}$$

The parameter analyzer has the provision for both parallel or series connection measurements. Typically, a series measurement circuit is used for low-impedance and parallel circuit for high-impedance samples [101]. The parallel connection model with $C_P - D_P$ has been chosen for the standard extraction of capacitance values. The dissipation factor D_P is calculated out of

$$D_P = \frac{G_P}{2\pi f C_P} \tag{3.6}$$

with G_P as conductance. An ideal capacitor has a dissipation factor of $D_P = 0$ $(G_P = 0)$, therefore $D = 0.1$ is the used limit for all $C(V)$ measurements in parallel equivalent circuit done in this work. Nevertheless, transformations between both parallel and series connection can be calculated according to Equation (3.7)

$$C_P = \frac{1}{1 + D_S^2} C_S \quad ; \quad G_P = \frac{D_S^2}{1 + D_S^2} \frac{1}{R_S} \tag{3.7}$$

Equal to experiments with current voltage measurements, a different degree of crystallinity and therefore different k-values in the dielectric layer have significant influence to results for capacitance voltage ($C(V)$) measurements of a MIM capacitor stack. Death layer effects [107, 108, 109] or interfacial layers arise out of varying permittivity perpendicular to the applied electrical field (capacitors with different permittivity in serial arrangement). Dielectric layers with different k-values are currently used in capacitor applications e.g. ZAZ (ZrO_2-Al_2O_3-ZrO_2) layer stacks (see Figure 1.2). With $C(V)$ measurements the dielectric permittivity can be extracted from the linear slope of CET versus thickness plots avoiding the interfacial layer.

More complex behavior appears for partially crystalline layers, where the dielectric permittivity changes in lateral dimension for the amorphous as well as the crystalline part. Neglecting any serial components like interface layers, $C(V)$ measurements of these capacitors result in a summation of capacitors with different k-value (Equation (3.8)).

$$\frac{C}{A} = \sum_i \frac{C_i}{A_i} = \sum_i \varepsilon_0 \, k_i \frac{1}{d} \tag{3.8}$$

Here, $\frac{C}{A}$ is the effective capacitance density of the complete capacitor with the size A and thickness d. $\frac{C_i}{A_i}$ is a variable proportion of parts with nanocrystallites as well as amorphous parts. With the knowledge of permittivities k_i for both parts, it is possible to calculate the crystalline proportions in the overall layer.

3.3.4 Relaxation measurements

The relaxation measurements have been performed using the similar setup as described earlier for the I(V) measurements. A voltage has been applied to the samples for 5 s similar to the delay time in I(V) measurements. When the applied voltage is turned off, the 'off' current is measured for 100 s in 0.1 s steps from the top electrode. The principle measurement scheme is shown in Figure 3.4. Due to a SMU dependent delay time t_d in milliseconds time frame, the exact starting time of measurement is different for all measured capacitors. Therefore only the relative time dependence of current has been compared in the samples.

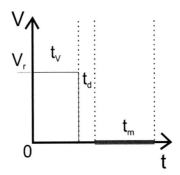

Figure 3.4: Principle measurement scheme for the extraction of relaxation currents. A voltage V_r is applied to the sample for t_V seconds. After turning off the voltage, a measurement unit dependent delay time t_d passes, before the time dependent current measurement starts (marked red). A measurement time t_m of 100 s with 0.1 s steps has been chosen to extract the decay of current.

4 Temperature stable high work function electrodes

4.1 Sputter deposition of electrodes

The complete metal stacks are sputtered in the UHV metal chamber on top of highly conductive Si substrates. Depending on the experiments to follow, substrates in the shape of full 150 mm wafers or 4×4 cm^2 Si pieces are used. The substrates have been baked out under vacuum at 150 °C to desorb H_2O molecules from the surface prior to deposition. An Ar^+ plasma treatment of the surface has been done to remove the native SiO_2 layer and other adsorbants from the Si substrates. The first layers as required for the bottom electrode stack consist of a conducting electrode material or a diffusion barrier (TiN or RuO_2). The standard sputtering conditions for diffusion barriers and or electrodes can be found in Section 3.1 in Table 3.1.

After deposition, some samples are treated with a post deposition anneal (PDA) *ex-situ* in a rapid thermal processing (RTP) furnace under atmospheric pressure with pure N_2 or an admixture of N_2 and 10% O_2 for TiN or RuO_2 respectively. This is done to increase the density and restore stoichiometry for TiN or RuO_2. Various temperatures have been tested for an annealing duration of 1 to 4 min. It was found, that the morphology of the layers does not change considerably after 1 min of annealing.

4.1.1 Titanium nitride layers

TiN layers are commonly used as electrode material in various industrial semiconductor devices. Their work function is higher (4.6 eV) [56] compared to that of various pure standard metal layers like Al (4.28 eV) or Ti(4.33 eV) [110]. For high band gap dielectrics like SiO_2, Si_3N_4, ZrO_2 and HfO_2, the conduction band offset is sufficiently high (>1 eV) in order to suppress leakage currents by thermionic emission. TiN is also known as a diffusion barrier for Pt [111]. This makes TiN the preferred electrode in beyond SiO_2

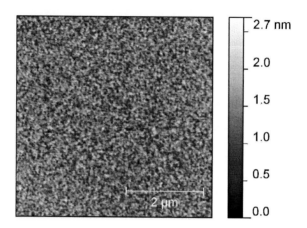

Figure 4.1: AFM topography map of an CVD deposited TiN layer (10 nm) on As-doped Si. The AFM image size is $5\times5\,\mu m^2$. The calculated roughness from the height profile is 0.3 nm (RMS).

device architecture. For high-k dielectrics with band gaps below 4 eV, even this work function may be too low, resulting in a Schottky barrier height (see Section 2.4.1) below 1 eV and therefore a higher leakage current due to Schottky emission for capacitors. TiN is discussed in this chapter as an electrode for comparative reasons, and moreover as a conductive diffusion barrier for noble metals (see Section 4.1.3). In this case, the roughness and temperature stability are the main attributes.

Various deposition technologies for TiN are available. In this work, the TiN layers are prepared with reactive sputtering from a Ti target using Ar and N_2 plasma. Before the introduction of a customized TiN deposition process, 300 mm CVD TiN covered highly As-doped Si wafers fabricated at Qimonda have been used (gas-phase doped Ar 1×10^{20} at/cm^3). In Figure 4.1, an AFM image of CVD TiN depicts the roughness of a TiN layer on top of a Si wafer. Wafers stored some time at ambient air exhibits a roughening of the surface due to the formation of TiON [112]. TiON is a low k dielectric and therefore an unwanted interfacial layer for a capacitor stack[113]. This is accompanied by an increase in roughness. In Figure 4.2, a cross-section TEM micrograph of a polycrystalline TiN layer with a thickness of 20 nm on top of a highly As-doped Si wafer is shown. Si is oriented along the (100) direction. A thin native SiO_2 oxide layer between the Si substrate and the TiN layer is visible.

The increase in roughness after the anneal of TiN in N_2 atmosphere (Figure 4.3(b))

Figure 4.2: TEM micrograph of CVD TiN on Si substrate. A native SiO_2 oxide layer is visible at the interface.

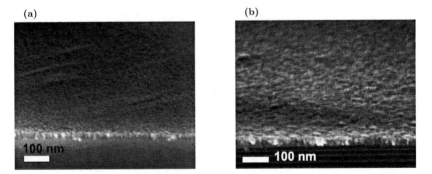

Figure 4.3: Tilted cross-section SEM images of TiN substrate (CVD) (a) before and (b) after anneal at 650 °C for 4 min in N_2 atmosphere. The surface morphology shows a slight increase in roughness after annealing.

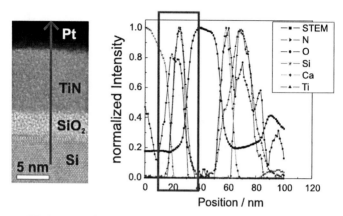

Figure 4.4: EELS linescan (STEM-HAADF mode) from a sputtered TiN/Pt electrode stack (HRTEM micrograph of the stack on left side). The highly doped Si wafer has not been precleaned with an Ar plasma prior to the deposition of the electrode, resulting in a 3 nm SiO$_2$ layer. The EELS linescan shows a significant amount of oxygen in the TiN layer (red rectangle).

can be a major issue for the scaling of thin CaTiO$_3$ capacitors. Therefore a TiN sputter process has been developed with the aim to reduce the roughness after annealing and further allow an *in-situ* deposition of the full capacitor stack without breaking vacuum. The TiN layers are sputtered on top of highly doped Si wafers at 300 °C as the compromise between surface roughness and TiN density. The flow rate of 4.5 sccm N$_2$ in 20 sccm Ar at a pressure of 1.6 × 10^{-3} mBar shows the optimal parameter set to ensure a high deposition rate and a stoichiometric TiN layer. The experiments result in TiN layers with significantly reduced roughness when heated up to 650 °C. TEM evaluation of CaTiO$_3$ capacitors with sputtered TiN layers reveal a measurable content of oxygen in the TiN layer shown by EELS line-scans of the complete capacitor stack (STEM-HAADF mode)[1] (Figure 4.4). An oxygen content has also been verified using XPS measurements. Nevertheless, the conductivity of sputtered TiN layers measured with a 4-point-probe setup is comparable to literature values of TiN layers and has not been altered by the oxygen content. Owing to the fact that sputtered TiN layers are mainly used as diffusion barrier for Pt (see end of this chapter), further attempts to reduce the oxygen content have been skipped.

---

[1]High Angle Annular Dark Field (HAADF). Higher atomic number areas appear brighter in the image.

(a)   (b)

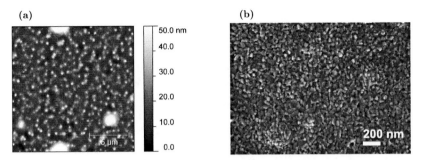

Figure 4.5: AFM topography map a) and SEM image b) of the Ru surface deposited at 400 °C on top of a Si substrate with native oxide. a) The AFM image shows large hillocks all over the surface. The calculated roughness is with 9.7 nm RMS very high. b) Beside these hillocks, the surface formation of approx. 50 nm diameter Ru crystallites is observed in the SEM image.

4.1.2 Ruthenium and Ruthenium dioxide layers

Ruthenium is a noble metal and belongs to the Platinum group. It has a work function of 4.6 eV, which is comparable to TiN. Conducting metal oxide electrodes like RuO_2 and $SrRuO_3$ are interesting for high-k oxides due to their higher work function of ~ 5.0 eV [48]. RuO_2 exhibits a small bulk resistivity ($40\,\mu\Omega$m), good thermal stability and diffusion barrier capability (see Table 2.1) [114]. As Ru, it can easily be etched with RIE (reactive ion etching) using O_2 as reactant under formation of $RuO_4 \uparrow$. Before the introduction of TiN as top electrode material, sputtered Ru has been used as top electrode for all capacitor stacks.

Sputtering metals including Ru at low temperatures result in very plane electrode surfaces. The sputter conditions for Ru deposition can be found in Section 2.2. When increasing the deposition temperature of the electrode to 400 °C, a formation of large crystallites (Figures 4.5(a) and 4.5(b)) begins. A formation of Ruthenium silicide Ru_2Si_3 has not been observed at this deposition temperature. According to literature, a silicide formation starts at 450°C [115, 116], whereas the native SiO_2 can inhibit this formation to higher (annealing) temperatures [117].

In Figure 4.6, the temperature dependent behavior of Ru on a $CaTiO_3$ layer describes a typical experiment using Ru electrodes. Although the heating process is done under high grade N_2 atmosphere, a reaction of Ru to RuO_2 starts at 505 °C and Ru is fully oxidized at 575 °C. The oxygen is most certainly taken from the underlying $CaTiO_3$ layer,

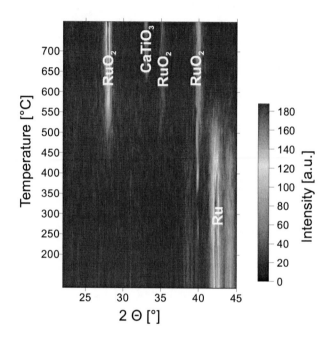

Figure 4.6: HTXRD of a Ru/CaTiO$_3$ stack on top of Si substrate. The sample was heated under N$_2$ atmosphere. The Ru layer nevertheless starts to oxidize at 505 °C to RuO$_2$ and is fully oxidized at 575 °C.

Figure 4.7: Time sequence of RuO$_2$ layer captured with microscope (50x). The violet colored layer on a Si substrate peels off immediately after contact to air.

because a high amount of adsorption from atmospheric gases can be excluded due to the elevated temperatures. This should significantly increase the amount of oxygen vacancies in CaTiO$_3$ and therefore increases the leakage current of the capacitor device. If used as a bottom electrode, the lateral expansion when oxidized to RuO$_2$ will destroy the capacitor structure above. These experiments conclude, that Ru is incapable for high temperature processes.

As alternative, the direct deposition of RuO$_2$ with the use of reactive sputtering has been tested. The oxide electrode should be stable at high temperatures and do not reduce the CaTiO$_3$ layer. Several RuO$_2$ sputter experiments were carried out showing the difficulty of the deposition. The sputtered layers are hygroscopic and quickly adsorb water from ambient atmosphere (Figure 4.7). The annealing of those RuO$_2$ layers in oxygen atmosphere prevented the further destruction of the surface. That indicates, that the sputtering process results in a understoichiometric oxygen proportion and therefore a hygroscopic RuO$_2$ layer. After anneal, the roughness of the layers is much larger compared to the TiN layers seen before. Further experiments for RuO$_2$ as electrode and diffusion barrier have been done by F. Silze as a student research project [118]. The results of the work underline the difficulty of stabilizing RuO$_2$.

As conclusion, Ru has been used as top electrode deposited at a low temperature of 100 °C, after high temperature processes were applied to the capacitor. With dry etching, the capacitor stacks were structured and measured electrically. The full noble metal electrode capacitor showed extremely low leakage currents (see Chapter 7 for details).

4.1.3 Platinum electrodes

Platinum has a the work function of 5.6 eV and a lattice constant comparable to CaTiO$_3$. These properties make Pt a formidable choice as electrode material in a CaTiO$_3$ capacitor stack. Nevertheless the use as electrode is hindered on Si as substrate, because Pt forms various Silicides (PtSi$_x$) at temperatures starting at 220 °C. The full reaction of Pt on

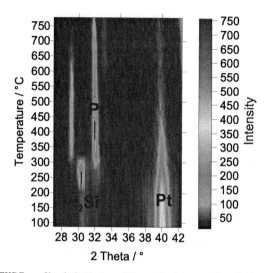

Figure 4.8: HTXRD profile of silicidation of Pt on Si substrate. The silicidation of Pt to Pt$_2$Si starts at 220 °C. At 310 °C, Pt$_2$Si further reacts to PtSi. The Pt signal decreases with the formation of silicides and at 450 °C, the process is finalized. The intensity of the (101) PtSi reflection at 29° decreases continuously starting at 600 °C, which hints towards a texture change in PtSi.

Si towards PtSi is shown in the HT-XRD spectrum in Figure 4.8. SEM images of the surface further depict the transformation of the Pt surface with the reaction of Si in Figures 4.9(a) and 4.9(b). For as-deposited Pt on Si, the Pt surface is very flat, as it is typical for sputtered metal layers. Annealing the sample to 650 °C shows a significant roughening due to the formation of PtSi$_x$. Also PtSi$_x$ layers are conductive and can be used as an electrode, the surface roughness prevents any use in ultrathin capacitor stacks. The usage of Pt as an electrode in high temperature processes, which are necessary for crystalline CaTiO$_3$ samples, therefore require a smooth diffusion barrier to prevent any formation of silicides.

4.1.4 Platinum on Titanium nitride

As mentioned previously, Pt needs a diffusion barrier for high temperature processes that prohibits the formation of platinum silicides. The formation of silicides start at 220 °C

(a)

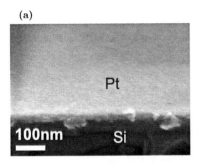

(b)

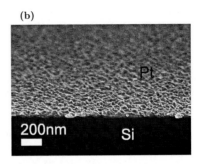

Figure 4.9: Tilted cross-section SEM images of the Pt on Si after deposition at room temperature (Figure 4.9(a)) and annealed at 650 °C in N_2 atmosphere for 4 min (Figure 4.9(b)). While the as-deposited Pt layer depicts a very smooth surface, the annealed sample shows a large increase in roughness due to the formation of Pt silicides.

(Figure 4.8). TiN offers an easy and low cost option as conductive diffusion layer between Pt and the Si substrate. The suitability of TiN in high temperature processes has been verified earlier in this chapter. The CVD grown TiN layers (Figure 4.3) have been tested in the electrode stack (Figure 4.10). The Pt layer is deposited on top of the TiN/Si substrate at 100 °C and shows a very smooth surface (Figure 4.10(a)). After annealing this stack at 650 °C in N_2 atmosphere, a slight increase in roughness can be observed in Figure 4.10(b). The Pt layer can not be distinguished from the TiN layer in the SEM cross-section. The sharp Si - electrode interface indicates that no silicide is formed on the surface. This is supported by XRD measurements, where no $PtSi_x$ reflection signal is visible in the spectrum. TiN is therefore a suitable diffusion barrier to prevent Si or Pt diffusion. Nevertheless, the visible roughness increase after anneal in Figure 4.10 makes it difficult to receive a uniform $CaTiO_3$ layer with thicknesses below 15 nm. For capacitors with the targeted oxide thickness and low leakage currents, a further reduction of electrode roughness is necessary.

To improve the roughness after high temperature treatments, a TiN process has been developed (Table 3.1 on Page 23). That allows the *in-situ* deposition of the complete electrode stack (Pt on TiN) without breaking vacuum. The TEM results of capacitors with both electrodes (Pt on CVD TiN vs. *in-situ* TiN/Pt), shown in Figure 4.11, prove the superiority of the *in-situ* deposition. Although the Si substrate exhibits a native oxide layer in both samples, the interface between TiN and Pt is significantly sharper

(a)

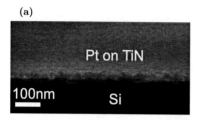

(b)

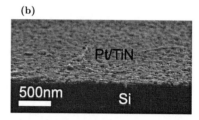

Figure 4.10: Tilted cross-section SEM images of Pt/TiN stacks on Si. (a) 20 nm Pt has been deposited on 20 nm TiN on a Si substrate at 100 °C. (b) The same sample annealed at 650 °C in N_2 atmosphere for 4 min shows an increased roughness. The Pt seems to mix with the TiN layer. Nevertheless no formation of Pt silicides is visible, the Si - electrode interface remains sharp.

for the *in-situ* process compared to the Pt on CVD TiN deposition. For that reason, the Pt surface is much smoother, allowing a further scaling of the full $CaTiO_3$ capacitor stack. Tests indicate, that even 3 nm TiN under a 40 nm Pt layer works as a suitable barrier towards Si, when deposition for several hours at 600 °C takes place. For stability and conducting reasons, a thicker TiN with 12 nm thickness has been chosen as standard diffusion barrier thickness for a 20 nm Pt metal layer.

4.2 CVD deposition of carbon electrodes

Carbon is an interesting electrode material with high conductivity and thermal stability [119], as it can replace TiN as electrode in the capacitor stack. It has been tested successfully in the integration in the deep trench DRAM front end process with several high-k dielectrics in Ref. [120].

In this work, carbon layers on Si wafers have been prepared in a CVD furnace at 860 °C for 10 min with the use of ethene (C_2H_4) at a pressure of 300 mbar. The recipe has been developed and described in detail in Refs. [121, 122]. A hot wall tube furnace (ATV PEO 603) has been used to create the pyrolytic conductive carbon electrodes. The samples were heated up and cooled down in nitrogen ambient. The deposition time was set to 10 min for a 110 nm thick layer. Silicon wafers with 200 mm diameter are used as substrates. Prior to the deposition of carbon, the native SiO_2 oxide has been removed with a hydrogen fluoride dip and put under vacuum within 7 minutes. A post-deposition anneal for 1 min at 1000°C in N_2 was done to remove the remaining hydrogen

(b)

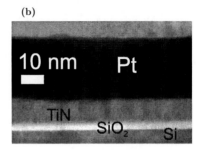

(a)

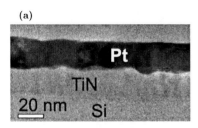

Figure 4.11: TEM micrographs of a CVD TiN/Pt (a) and an *in-situ* sputtered TiN/Pt (b) stack. Both samples are heated to 600 °C during the deposition of CaTiO$_3$ for 120 min and 40 min respectively. The interface Pt/TiN is much rougher for the CVD TiN sample (a), resulting in a rougher Pt surface compared to the *in-situ* TiN/Pt electrode stack (b).

contamination.

The prepared carbon layers exhibit an astonishing surface uniformity. A SEM cross-section image of a prepared sample is shown in Figure 4.12. The surface remains planar after the 1000°C anneal, which makes the carbon layer a suitable electrode for the high temperature deposition of CaTiO$_3$. Other comparable images of carbon layers can be found in Refs. [121, 122].

4.3 Conclusion

This chapter deals with the preparation and characterization of electrodes for implementation in ultrathin metal-CaTiO$_3$-metal capacitors. The electrodes have been optimized for roughness, temperature stability and possible oxidation. Electrodes with different work functions and interface properties are needed to elucidate the properties of CaTiO$_3$. The electrodes are optimized to yield low surface roughness in order to reduce field enhanced leakage currents. Different diffusion barriers are evaluated for suitability for the integration of noble metal electrodes. A temperature stable electrode with Pt as electrode material on TiN as diffusion barrier was found to be the best choice regarding work function. roughness and temperature stability. Pure sputtered TiN layers show a smooth surface after anneal but lack for stability in oxygen environments (e.g. during CaTiO$_3$ sputtering). CVD grown carbon electrodes exhibit the smoothest surface after

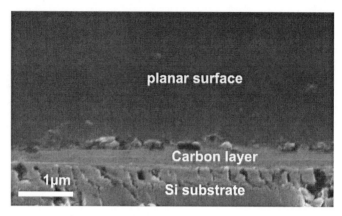

Figure 4.12: Tilted cross-section SEM image of an annealed carbon layer on Si. The surface remains planar even with annealing temperatures of 1000°C. The layer thickness is indicated in between the orange lines.

annealing at 1000°C. The suitability of carbon as a material with sufficient band offset towards CaTiO$_3$ will be tested later in Chapter 7.

5 Physical investigation of calcium titanate layers

5.1 Experiments on optical properties and density of CaTiO$_3$ layers

Optical methods give a convenient and nondestructive access to some important properties of CaTiO$_3$ layers like refractive index and band gap in the accessible wavelength range. In literature, various experiments had already been conducted for thin CaTiO$_3$ layers on various substrates (see Section 2.1.2 for details). For different applications, various deposition techniques are required, resulting in differences of CaTiO$_3$ properties. As an example, sputtered layers are known to be oxygen deficient, e.g. see Ref. [123]. The oxygen deficiencies influence the optical properties as well. The results of optical measurements help to classify the sputter deposition process for CaTiO$_3$ layer properties and allow the comparison to results from literature.

5.1.1 Refractive index and extinction coefficient of CaTiO$_3$

The use of spectroscopic ellipsometry in the visible and ultraviolet (UV-vis) (250 nm to 800 nm wavelength) range provides access to the layer thickness and the dispersion values of CaTiO$_3$. By employing applicative models to the dispersion dependence on wavelength, some statements about the optical band gap are possible. The explanation for the extraction of optical constants as well as the band gap can be found in Chapter A. The fit with the Tauc-Lorentz model to extract refractive index and extinction coefficient shows the best agreement with experimental data using a band gap of 3.97 eV for the amorphous layer. In Figure 5.1a), the refractive index for amorphous and partially crystallized CaTiO$_3$ is compared to values from literature. At a wavelength of 632 nm, the extracted refractive index is lower compared to most literature values. Only results of sputtered layers with a low density of 3.0 g/cm^3 show a comparable refractive index at

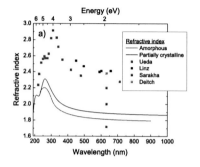 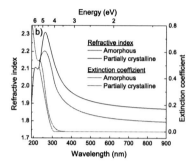

Figure 5.1: Dispersion values of sputtered $CaTiO_3$ layers. a) The refractive indices of $CaTiO_3$ in this work are lower compared to most literature values. This may be attributed to the low density of the sputtered $CaTiO_3$ films. b) When going to higher deposition temperatures, partially crystallized layers show a larger refractive index in the measured spectral range. The maximum lies at higher energies compared to measurements of Ueda et al. [12]. This can be attributed to the different measurement techniques (vacuum ultraviolet reflectometry versus ellipsometry [12, 15, 124, 125]).

the selected wavelength [125]. The other results from literature were obtained for bulk $CaTiO_3$ or layers annealed up to $900\,^{\circ}C$. This variation can be related to a denser film in a bulk crystal or after annealing. In sputtered layers, Ar is typically incorporated into the layer, which reduces the overall density. In addition, the results vary for different experimental setups (ellipsometry, reflectometry, absorption spectroscopy).

5.1.2 Band gap determination with low energy electron energy loss spectroscopy

The previously used ellipsometry required models to extract the band gap of $CaTiO_3$. The results obtained this way are fully dependent on the used model and do not necessarily describe real physical constants. Another disadvantage of optical measurements is the access only for direct transitions. In contrast, accelerated electrons have a large impulse, which can compensate the low probability of photon-phonon interactions for photons necessary for indirect transitions.

Low energy electron energy-loss spectroscopy (LE-EELS) with TEM allows another ac-

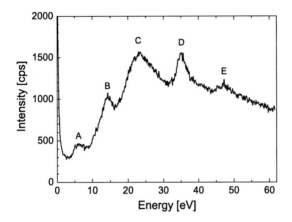

Figure 5.2: EELS spectrum of $CaTiO_3$ in the range from 0 eV to 62 eV. It shows the lowest band transition at Point A as well as other excitations from electron beam (STEM, 80 kV acceleration voltage, monochromatic electron beam, FWHM = 0.18 eV). The energies of Peaks A-F are given in Table 5.1.

Table 5.1: Peak/edge positions (in eV) of the energy loss function of a $CaTiO_3$ lamella as labeled in Figure 5.2. The suggested assignments are given in parenthesis. Above 20 eV, various valence band - conduction band as well as inter-band transitions are possible. Peaks D and E correspond to shallow core level transitions of Ca and Ti respectively.

	A (eV)	B (eV)	C (eV)	D (eV)	E (eV)
$CaTiO_3$ (this work)	6.0	14.4	23.4	35.2	47.3
$CaTiO_3$ [12]	6	15	24		
$SrTiO_3$ [126]	6.4	13.3	-	37.5	45.5
Assignments	O 2p - Ti 3d	(O 2p - Ca 4s)	(Ca 4s - Ti 3d)	(Ca M_1 edge, ..)	(Ti $M_{2,3}$ edge, ..)

cess to the band gap of the bulk material including both direct and indirect transitions. This technique measures the energetic loss of electrons transmitted through a $CaTiO_3$ lamella. With a much larger momentum of electrons than equivalent photons, both the energy and momentum are transferred to crystal electrons. This allows the observation of both direct and indirect transitions. The interaction of electrons with solids have often been investigated in literature, a full derivation of the intensity dependencies for TEM via JDOS (joined density of states) theory can be found in Ref. [127]. The energy loss of electrons for a direct transition in the solid follows Equation (5.1), with E as the energy loss of a single electron and E_g^{direct} as the direct band gap of the solid.

$$Intensity \propto (E - E_g^{direct})^{1/2} \qquad (5.1)$$

For indirect transitions, the intensity depends on $E_g^{indirect}$ as in Equation (5.2).

$$Intensity \propto (E - E_g^{indirect})^{3/2} \qquad (5.2)$$

The electrons are inelastically scattered at atom shells, when the energy loss reaches the energy level of the band gap as minimum. Electrons showing a signal below the band gap of the solid are typically part of the Gaussian shaped zero-loss electrons or electrons losing energy in form of Čherenkov radiation, when electrons faster than the phase velocity of light in the solid are slowed down. The first energy loss of electrons is measured, when incident electrons excite crystal electrons from the valence band or other occupied orbitals to unoccupied orbitals. With this method, a direct observation of the band gap (direct or indirect) of an insulator is possible [127, 128, 129]. Beside band-to-band transitions, the incident electrons excite surface and bulk plasmons typically at energy levels around 10-20 eV, which also contributes to an energy loss.

For $CaTiO_3$, the band edge from a LE-EELS measurement is shown in Figure 5.3. According to Equations (5.1) and (5.2), a direct band gap has been extracted with $E_g^{direct} = 4.38$ eV as well as the indirect band gap of $E_g^{indirect} = 3.83$ eV. This result correlate well with recent results from simulations of $CaTiO_3$ [9, 20]. The extracted band gap of 3.8-4.38 eV is slightly higher then other EELS measurements of $CaTiO_3$ from *Ueda et al.* [13]. The difference is attributed to a better energy resolution with the usage of a monochromatic electron beam and a reduced acceleration voltage of 80 kV in this work. These settings reduce the Čherenkov radiation and plasmonic excitations and result in a better signal-to-noise ratio near the zero energy peak. The difference of $E_g^{indirect}$ and E_g^{direct} of $CaTiO_3$ is identical to measurements with EELS and vacuum ultraviolet (VUV) spectroscopy for $SrTiO_3$ with 3.25 eV and 3.75 eV, respectively [126, 130].

Various simulations on nanosized high-k materials reveal, that the band gap of a dielectric is reduced, when reaching the border of each crystal grain [21, 131]. The reduction

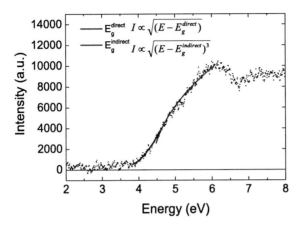

Figure 5.3: Low energy EELS measurement of polycrystalline $CaTiO_3$ in a MIM capacitor stack. (STEM, 80 kV acceleration voltage, monochromatic electron beam, FWHM $= 0.18\,\mathrm{eV}$). The band gaps are extracted for $E_g^{indirect} = 3.83\,\mathrm{eV}$ (blue line) and $E_g^{direct} = 4.38\,\mathrm{eV}$ (red line) respectively. These values do not change for measurements across grains, planar or perpendicular to the $CaTiO_3$ layer in the resolution range for STEM (Figure 5.4).

arises from surface relaxations of atoms and induce shifts of both valence and conduction bands. Additionally, dangling bonds and a high concentration of oxygen deficiencies at grain boundaries form trap bands below the conduction band.

Different EELS scans perpendicular and in plane of the $CaTiO_3$ layer have not shown a change in band gap across interfaces and grain boundaries, which can be associated to the former described band gap change. Nevertheless the intensity change of a line scan in plane of the $CaTiO_3$ layer in Figure 5.4 can be identified as different grains. The decrease of intensity of the electron beam is the result of electron scattering along grain boundaries. No significant change of the band gap of $CaTiO_3$ is visible from EELS, when reaching a grain boundary. Either the energy level of any reduction in band gap is smaller than the energy resolution of the EELS measurement, or the local excitation coil of the electron beam is much larger compared to the very localized grain boundary. Simulations revealed a change of band gaps up to a thickness of 1-2 monolayers.

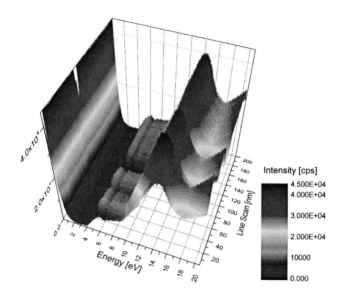

Figure 5.4: 3D plot of a 200 nm EELS line scan in a 35 nm CaTiO$_3$ layer between metal electrodes in an energy range between 0-20 eV. The band gap as extracted in Figure 5.3 shows no measurable change along the measured scan line. The layer is fully crystalline (see TEM micrographs in Chapter 6 for details). The change in intensity along the line scan can be attributed to scattering of electrons along grain boundaries.

Table 5.2: Parameters for fitting the XRR profiles in Figure 5.5

Sample	Thickness (nm)	Rel. density of $CaTiO_3$	Roughness (nm)
black	51.4	0.80	0.66
red	31	0.90	0.4
green	31.6	0.93	1.11
blue	51.4	0.93	0.47

5.1.3 Density approximation with X-Ray reflectance

From X-Ray reflectance (XRR) measurements (usually 2θ from 0° to 6°), thickness and density of the deposited films shall complement the results from ellipsometry. The analysis of XRR measurements needs to set up a model of the layer structure and fit several parameters including the thickness to the experiment. Therefore, a single layer of $CaTiO_3$ has been deposited at typical deposition temperatures on top of a Si substrate (with native oxide) to allow a convenient fit of the thickness, density and roughness. The thickness extracted from XRR measurements are in agreement to the ellipsometric data. This shows that the used Tauc-Lorentz model for the measured ellipsometry data give reasonable results of physical properties of $CaTiO_3$. One can further concentrate on the density, which influences the dispersion values significantly (see Figure 5.1b)).

The density in XRR measurements is derived from the critical angle (θ_c), where an abrupt fall in reflected intensity occurs. θ_c and the shape of the total reflection edge is the result of an interaction between both the surface density and roughness [132]. In Figure 5.5, the intensity over 2θ from various $CaTiO_3$ layers is depicted. Table 5.2 summarizes the parameters used to fit the XRR profiles. The lowest density derived from XRR is the $CaTiO_3$ layer deposited at 300°C, the amorphous layer shows the lowest density. With increasing the deposition temperature above the crystallization temperature of $CaTiO_3$, the fitted density increases as well. This is in agreement with the dispersion values from ellipsometric measurements done in this work and to literature. One has to keep in mind that a high interface roughness typically correlates with lower densities, which is not observed here.

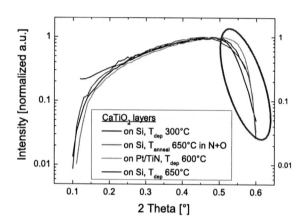

Figure 5.5: X-ray reflectance of various layers at typical deposition temperatures. The edge of the signal describes the density of the different layers. The most dense $CaTiO_3$ layer is deposited above its crystallization temperature on Pt electrode layer (green line). The least dense $CaTiO_3$ layer has been deposited on Si at 300 °C (amorphous, black line). After annealing the sample at 650 °C in N_2-O_2 atmosphere, the layers roughness increases and shifts the maximum intensity to a lower angle with increasing density (broader peak, red line). The thickest sample at highest deposition temperature of 650 °C on Si (blue line) shows a density between the annealed sample and the sample on top of Pt.

Table 5.3: Literature values for binding energies of Ca, Ti, O in CaO, TiO$_2$ and CaTiO$_3$.

Material	Ca $2p_{3/2}$ (eV)	Ti $2p_{3/2}$ (eV)	O $1s$ (eV)
TiO$_2$		459.0[a] 458.8[b]	
CaO	347.2[b] 346.6[c]		531.2[c]
CaTiO$_3$	346.6[b] 347.6[d]	458.6[a] 458.4[b]	531.3[d]

[a] Murata et al. [133] ; [b] Hanawa et al. [134] ; [c] Demri et al. [135] ; [d] Asami et al. [136]

5.2 X-ray photoelectron spectroscopy of CaTiO$_3$ layers

The investigation of surface and bulk binding states of CaTiO$_3$ can be done using X-ray photoelectron spectroscopy (XPS). The results from this measurement give answers to chemical environment, binding energy and oxidation states of elements contained in the layer. The high resolution allows the observation of atom concentrations down to 1 % of the bulk material. This includes dopants or unwanted contamination (e.g. carbon) of the layer above that resolution limit.

As for CaTiO$_3$, following questions shall be resolved:

1. Is Ti at its full oxidation state (Ti^{4+}) or is any significant composition with a lower oxidation state (oxygen vacancies) measurable in the layer (typical for sputtered layers)?

2. Is there a layer of Ca(OH)$_2$ and CaCO$_3$ on the surface of CaTiO$_3$ due to the hygroscopic nature of Ca?

3. Are there contaminations of foreign atoms (e.g. Fe, F, Ni, Mo), which may influence the electric conductivity and the formation of trap states?

4. What is the measured concentration of Ca, Ti and O and is it in the stoichiometric range of CaTiO$_3$?

Literature values for binding energies of Ca,Ti,O from the CaO, TiO$_2$ and CaTiO$_3$ are summarized in Table 5.3.

XPS experiments of amorphous CaTiO$_3$ stacks with and without a Ru capping layer have been performed and the binding levels of all elements in the available energy range have been extracted. The comparability to values from bulk or crystalline CaTiO$_3$ from literature [133, 134, 135, 136] is obtained. From pair/radial distribution function results [137], it is known, that the local order does not change between amorphous and crystalline samples respectively [138, 139]. Therefore, XPS should not distinguish between

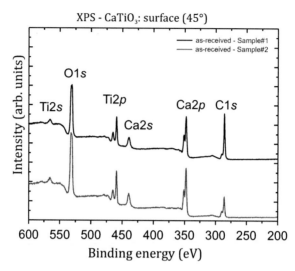

Figure 5.6: XPS spectrum of a $CaTiO_3$ stack between 200 eV to 600 eV. The two samples show no difference in composition. Small amounts of C and F reveal the adsorption of carbon dioxide as well as contamination from sample boxes.

amorphous and crystalline layers, when the binding length/energy is constant. Any shift is finally attributed to a chemical change in the layer.

An overview of energy levels for $CaTiO_3$ in the range from 200 eV to 600 eV is depicted in Figure 5.6, and contains all relevant data for Ca, Ti and O. With an incidence angle of 45°, both surface and bulk properties are measured. The different binding energies of Ti, Ca and O from as deposited (Figure 5.7) as well as sputter cleaned samples (Figure 5.8) are now investigated in detail.

The Ti binding states (Figure 5.7b)) from as deposited layers as well as sputtered layers after removal of contaminants (Figure 5.8b)) reveal a strong Ti^{4+} peak from TiO_2 bindings. A small shoulder from a Ti^{3+} signal in as deposited layers, which is enhanced significantly when sputtered, indicates the existence of small amounts of Ti_2O_3 or oxygen vacancies [140]. This result gives a hint towards an oxygen deficiency from the deposition process of $CaTiO_3$. The binding energy of 459.2 eV from Ti^{4+} ions is higher as from powder samples in $CaTiO_3$ [133, 136] and slightly higher in TiO_2 [133]. The binding energy of Ti^{4+} is dependent on the binding length between the Ti - O binding in the TiO_6 octahedron and typically shifted to lower values with introduction to bivalent cations (Ca,

Sr, Ba; from left to right higher to lower binding energy)[141]. The usage of ultrathin CaTiO₃ samples in this work may lead to suppression of vibrational modes responsible for the binding energy position [141]. The energy separation between the two spin-orbit components Ti $2p_{3/2}$ and Ti $2p_{5/2}$ of 5.7 eV also indicates the presence of Ti in the Ti^{4+} oxidation state [142].

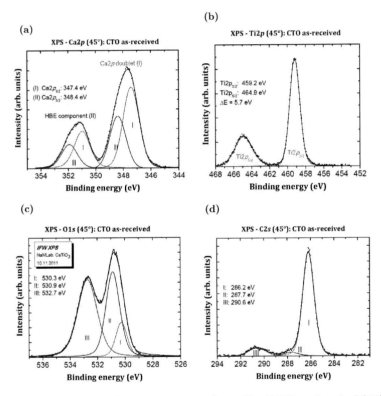

Figure 5.7: X-ray Photoelectron spectroscopy of crystalline CaTiO₃ on top of a Si(100) substrate. The binding energy of the elements a) Ca, b) Ti, c) O and d) C are magnified from Figure 5.6. The oxygen characteristic peaks correspond to : I - O^{2-}(TiO₂); II - O^{2-}(CaO, CaTiO₃), (CO₃)$^{2-}$, OH$^-$; III - H₂O according to Refs. [134, 143].

The XPS spectrum of binding energies for the Ca doublet (Ca $2p_{3/2}$ and Ca $2p_{1/2}$) is

5 *Physical investigation of calcium titanate layers*

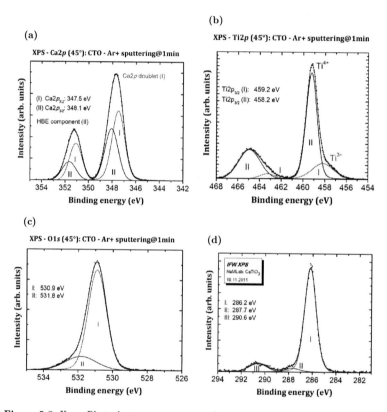

Figure 5.8: X-ray Photoelectron spectroscopy of crystalline CaTiO₃ on top of a Si(100) substrate after sputtering with Ar to remove surface adsorbents. The spectra correspond to the electron energy of a) Ca, b) Ti, c) O and d) C. For O, the peak
corresponding to the adsorption of H₂O vanished completely.

given in Figures 5.7a) and 5.8a), respectively. Both wide peaks indicate the existence of a sum of at least 2 doublets of Ca in binding (here components I and II). A Ca doublet from $CaTiO_3$ at 347.4 eV and 350.9 eV is accompanied with a doublet from another high binding energy (HBE) component at higher energy of 348.4 eV and 351.9 eV. The HBE component can be assigned to a Ca-O binding in CaO or $Ca(OH)_2$[1]. The HBE peaks suggest the existence of CaO or its hydroxide and disappear after surface cleaning. The FWHM of the peaks is lowered for the sputtered layers, the HBE component peak is reduced and only Ca-O bindings from $CaTiO_3$ remain. The binding energy of Ca in $CaTiO_3$ corresponds to other $CaTiO_3$ results obtained by annealing $CaTiO_3$ in vacuum [136] and is slightly higher than in CaO [134] (Table 5.3). This indicates, that the Ca-O binding in $CaTiO_3$ is not significantly influenced by the potential field of the neighboring atoms.

The most significant changes from XPS spectra between as-deposited and plasma cleaned $CaTiO_3$ can be seen in Figures 5.7c) and 5.8c). Three peaks can be assigned in the XPS spectrum of Figure 5.7c) which are referred as adsorption of water and some molecules containing carbon. Interestingly, the small shoulder (Peak I) at lower energy is attributed to the presence of pure TiO_2. While the CaO and $CaTiO_3$ peaks of oxygen can not be distinguished from each other in Peak II, Peak I indicates a decomposition of $CaTiO_3$ to CaO and TiO_2 followed by the formation of $Ca(OH)_2$. This is the first hint towards the reason for hygroscopy in $CaTiO_3$. After plasma treatment, the Peak III, which indicates the existence of H_2O, is completely removed and only a single Peak at 530.9 eV remains (Peak I). The shoulder at higher energies is attributed to oxygen deficiency in $CaTiO_3$ (see Ti^{3+} formation in Figure 5.8b)) and also appears for TiO_2 together with higher concentrations of Ti^{3+} in Ref. [144]. The adsorption of carbon on the surface can be seen in Figure 5.7 d) and is removed after Ar plasma cleaning in Figure 5.8 d).

5.3 Stoichiometry of CaTiO₃ layers

XPS measurements show that the Ca to Ti relation exhibit identical concentrations as expected from sputtering from a stoichiometric $CaTiO_3$ target. Nevertheless, a strong oxygen deficiency is revealed. The composition of the as-received $CaTiO_3$ films could be determined from the XPS measurements to be $Ca_{1.5}Ti_{1.5}O_2$[2]. Every second oxygen atom is missing or not bound to Ca or Ti in the amorphous layer. However, extremely low

---

[1]NIST X-ray Photoelectron Spectroscopy Database 20, Version 3.5: srdata.nist.gov/xps/

[2]No surface oxygen annealing to restore oxygen stoichiometry has been done to keep $CaTiO_3$ amorphous.

leakage currents for amorphous $CaTiO_3$ layers (see Chapter 7 for details) are indicating a much smaller deficiency, no major contribution of conduction through trap states or even metallic behavior of unbounded metal atoms is measured. In addition, X-Ray fluorescence analysis (XFA)[3], where a much smaller oxygen deficiency has been measured, can not confirm the strong oxygen efficiency. The relative error of this specific setup for oxygen and lighter elements is approx. 15 %.

Rutherford backscattering (RBS)[4] on first sputtered $CaTiO_3$ layers revealed a Ca/Ti ratio of about 1.30-1.39. XPS measurements showed a ratio of 1.27 of the same as deposited $CaTiO_3$ layers. After extensive sputtering from the $CaTiO_3$ target, XFA scans revealed a Ca/Ti ratio of 0.98-1.03, which has been confirmed for other samples with XPS (see XPS results).

5.4 Roughness and crystallinity of $CaTiO_3$ layers

5.4.1 Surface roughness

The advantage of the sputter deposition process is the high uniformity and small roughness for as deposited films even on large scale substrates. The following AFM images of $CaTiO_3$ layers on top of an electrode shall be representative for complete capacitor stacks except the omitted top electrode. Most influence to the surface roughness is expected to arise from high temperature deposition combined with long deposition times.

Two extremes for surface roughness of $CaTiO_3$ layers during deposition are shown in Figure 5.9. The roughness value is significantly lower than 1 nm, even for the thick and crystallized layer (60 nm, high T_{dep} of 650°C) of $CaTiO_3$. Amorphous layers are typically very smooth compared to crystalline samples. As indicated in XRR measurements (Figure 5.5), the density of all layers do not change significantly. For amorphous layers, the smoothest surface is achieved between 300-450°C deposition temperature. With the further increase of T_{dep} to approx. 600°C (see Figure 5.5) the density but also the roughness increases. This may be attributed to a roughening of the underlying electrode. Above the crystallization temperature, the $CaTiO_3$ surface shows an increased roughness compared to the previous amorphous samples. No cracks or the formation of large crystals can be identified in the AFM image of Figure 5.9 b).

An annealing series of an amorphous $CaTiO_3$ layer is shown in Figure 5.10. Besides the formation of hillocks (brighter area in the image), the roughness is not changing

---

[3]The XFA measurements were done by Dr. Manfred Schuster at Siemens Analytics (CT T MHM ANA)
[4]Done by Dr. rer. nat. Steffen Teichert (Principal Surface and Materials Analysis, Qimonda/Infineon).

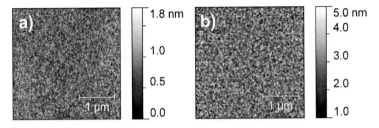

Figure 5.9: AFM image of an a) amorphous CaTiO$_3$ layer (30 nm, RMS 0.20 nm) compared to a sample b) deposited at 550°C (60 nm, RMS 0.47 nm) on top of Pt/TiN electrodes. Even at elevated temperatures for long deposition times, the roughness of CaTiO$_3$ increases insignificantly.

significantly for annealing times between 1-8 min. Figure 5.10 b), cracks and holes are developing, revealing a surface reconstruction compared to the flat amorphous sample. XRR measurements (Figure 5.5) showed no significant increase in density of CaTiO$_3$ for annealed samples, therefore the layer morphology may mainly change due to transition from amorphous to crystalline CaTiO$_3$ including the reduction of incorporated defects from Ar or oxygen vacancies [123]. Only insignificant change in roughness of the CaTiO$_3$ layer deposited at 650°C is expected, when an anneal step is applied afterwards.

In summary, the CaTiO$_3$ layers deposited at various temperatures show smooth and fully closed surfaces with an RMS < 1.0 nm in AFM measurements. The *in-situ* deposition of Ru as top electrode on those layers without any temperature post-treatment produces capacitors with low leakage currents (See Chapter 7). Post-deposition annealing of pure CaTiO$_3$ films reveals a development of holes and cracks for annealing times larger 2 min. Therefore any further electrode deposition for electrical characterization is of no use. No cracks in the surface have been observed after annealing of a complete capacitor stack including the TiN top electrode. These capacitors have also been processed further for electrical characterization (Section 3.1).

5.4.2 Crystallization of CaTiO$_3$

(Poly-)Crystalline CaTiO$_3$ layers are formed, when deposition or annealing temperatures increase above the crystallization temperature T_{cryst} of CaTiO$_3$. The crystallization is necessary to achieve the permittivities needed in MIM capacitor stacks. According to *Holliday et al.* [46], CaTiO$_3$ possesses a substrate dependent crystallization tempera-

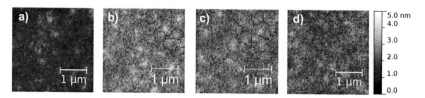

Figure 5.10: Annealing series from the amorphous layer of Figure 5.9 a) with a RTP anneal at 650°C in N_2 for a) 1 min (RMS: 0.38 nm) b) 2 min (RMS: 0.51 nm) c) 4 min (RMS: 0.50 nm) and d) 8 min (RMS: 0.41 nm). Larger hillocks appear after 1 min annealing with a diameter of about 80 nm. Holes and cracks (some are redrawn) appear on the surface visible in Figures b)-d), indicating a volume reduction of the sputtered layers. Focusing on the hillocks, the rearrangement of the surface is already finished after 1 min annealing time.

ture. On a Si substrate, the crystallization temperature is 650 °C. Temperature stable substrates/electrodes are needed to avoid damage to the MIM stack during deposition or annealing. Suitable electrodes have been proposed previously in Chapter 4 for that requirement.

Figure 5.11 shows a high temperature XRD (HTXRD) spectrum (Section 3.2) of a $Ru/CaTiO_3$ layer stack on top of a Si substrate with a native oxide. Both $CaTiO_3$ and Ru have been sputtered at room temperature. To investigate the crystallization of $CaTiO_3$, the sample has been heated in N_2 atmosphere. Besides the oxidation of Ru to RuO_2, the HTXRD spectrum shows a starting crystallization of $CaTiO_3$ at 640 °C. This is in agreement with the literature values for $CaTiO_3$ [46]. The Ru capping layer in this experiment protected $CaTiO_3$ from the adsorption of ambient gases and water during sample transportation. It may introduce stress to the dielectric during crystallization and therefore increases T_{cryst} of $CaTiO_3$. Nevertheless, the $CaTiO_3$ film thickness of 30 nm in this experiment is expected to be sufficient to neglect a distinct influence to the crystallization behavior.

With GIXRD on a MIM capacitor stack and a $CaTiO_3$ deposition temperature of 550°C (Figure 5.12), a crystal peak at $2\theta = 32.8°$ can be assigned to $CaTiO_3$. The other $CaTiO_3$ peaks are superimposed on the different electrode signals. This measurement depicts the difficulty to extract useful data for crystallization of $CaTiO_3$ in a full capacitor stack. Strong peaks are visible for Ru and Pt positions, which hide the other prominent peaks of $CaTiO_3$. Therefore, no reliable crystal size extraction (with the Debye-Scherrer-Method) is possible.

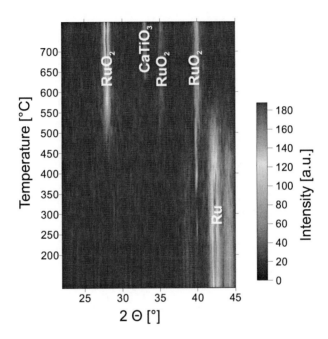

Figure 5.11: High Temperature XRD (HTXRD) spectrum of CaTiO$_3$ on top of a Si substrate with native oxide. A Ru top layer works as a capping layer for CaTiO$_3$. At room temperature, crystalline Ru reflections are visible, the CaTiO$_3$ layer is amorphous. RuO$_2$ formation starts at 505 °C, the Ru is completely oxidized at 575 °C. The crystallization of CaTiO$_3$ starts at 640 °C.

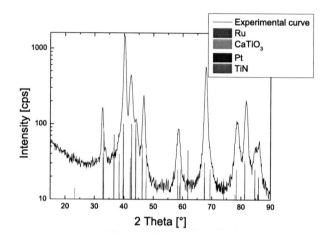

Figure 5.12: GIXRD of a Si/TiN/Pt/CaTiO$_3$/Ru MIM capacitor stack. CaTiO$_3$ has been deposited at 550 °C. Besides a peak at $2\theta = 32.8$ °, which can be associated with CaTiO$_3$, the metal electrode materials contribute to the spectrum and mask other potential peaks of CaTiO$_3$.

5.5 Conclusion and summary

The dielectric layers of CaTiO$_3$ have been investigated regarding the optical properties of the dispersion values, which are dependent on layer thickness, density and band gap. The dispersion values are lower compared to literature values of CaTiO$_3$ bulk crystals. On the one hand, this is explained by the lower density due to the sputter deposition of CaTiO$_3$. On the other hand, the extracted band gap of more than 4 eV is higher than in other publications and therefore the absorption edge shifts to higher energy values. This, beside other reasons, can be attributed to the high purity of the used CaTiO$_3$ sputter target (99.999% grade).

XPS measurements revealed the existence of adsorbents at the CaTiO$_3$ surface corresponding to CaO, Ca(OH)$_2$, CaCO$_3$ formation. This result confirms the hygroscopic nature of CaTiO$_3$ and a reaction with ambient gases. By pre-cleaning the surface in vacuum with the help of Ar$^+$ plasma, carbon content is removed significantly, as well as the other Ca containing components. The Ar$^+$ cleaning shows a reduction of Ti^{4+} to Ti^{3+} due to the preferential removal of oxygen in conjunction with the oxidation state change of Ti. The comparison of stoichiometry with EELS and XPS reveals a Ca/Ti ratio of 1 and a significant oxygen deficiency in the layers. Although sputtering is known to create layers with some oxygen deficiencies, the level measured especially with XPS has to be taken with great caution.

The oxide layer thickness is a crucial parameter, when thickness dependent information like permittivity has to be extracted. Therefore, the $CaTiO_3$ thickness calibration is done in this work with ellipsometry as well as XXR measurements and also backed with several TEM images. All methods provided similar thickness results, therefore ellipsometry as the most simple method has been extensively used to calibrate the thickness of $CaTiO_3$ oxide layers.

With the help of HTXRD measurements, the crystallization temperature of amorphous $CaTiO_3$ in the investigated layers lies at 640 °C. Rapid thermal annealing at 600°C of some samples as well as depositing samples at 550°C also induces crystallization. The crystallization temperature can be slightly influenced by the choice of substrates (Pt vs. Si).

6 Transmission electron microscopy characterization on crystallinity of calcium titanate

6.1 Crystallization behavior at 550°C

For various ABO_3 perovskites, the crystallization temperature lies above 600 °C when using typical deposition techniques (Sputtering, Pulsed laser ablation etc. of $SrZrO_3$ [145], $SrTiO_3$ [146, 147]). As depicted previously in Figure 5.11, this is also valid for $CaTiO_3$ deposited on Si substrates. In most publications, post deposition anneals (PDA) are preferred to achieve crystallinity due to better environmental control (gases, temperature, annealing time) and less temperature induced stress towards the substrate. PDA requires temperature stable top and bottom electrodes (e.g. TiN) when annealing the full MIM capacitor stack. As seen in Chapter 4, Ru starts to oxidize at 400°C and is not suitable as electrode material for the discussed annealing temperatures.

When using an reactive top electrode, e.g. Ru, the crystallization of the oxide layer is done before the deposition of the top electrode. In this case, the crystallization of the dielectric is not influenced or inhibited by lateral strain from the top electrode. In this work, for capacitors containing Ru top electrodes, the $CaTiO_3$ has been deposited at temperatures above 600 °C to achieve crystallinity. With TiN top electrode, PDA has been applied after the deposition of the full capacitor stack.

Both high temperature processes require bottom electrodes, which are stable during the complete deposition/annealing process. This is especially important for high temperature deposition processes, as the deposition time is much longer compared to the annealing time with PDA. Carbon or TiN/Pt as bottom electrodes have been selected in combination with $CaTiO_3$ to serve that purpose (see Section 2.2).

For a $CaTiO_3$ layer with a thickness of 35 nm deposited at 550 °C, a complete crystallization is observed in TEM (Figure 6.1). This temperature is lower than the results

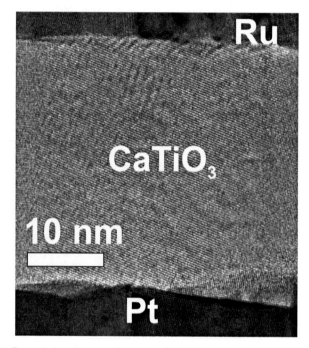

Figure 6.1: Transmission electron micrograph of CaTiO₃ layer deposited at 550 °C. The full layer is crystallized. No interfacial layer is visible between CaTiO₃ and the electrodes.

extracted from X-ray diffraction measurements in Figure 5.11. The difference in crystallization temperature T_{crys} between the samples can be summarized as follows:

- Annealing of a layer is perceptibly different to a deposition at high temperatures due to the different temperature budgets.

- With a Ru layer covering the $CaTiO_3$, strain was introduced to $CaTiO_3$ and therefore the crystallization may be inhibited in Figure 5.11.

- Sputtering at elevated substrate temperatures in conjunction with energetic material particles (typically $10\,eV$) may introduce enough energy for the formation of crystallites.

- Pt has has a comparable lattice constant to $CaTiO_3$. This reduces the energy necessary for crystallization, when using Pt as bottom electrode.

The effect of strain introduced to the $CaTiO_3$ layer as well as the influence of the Pt lattice match to $CaTiO_3$ is further investigated in the next section.

6.2 Strain in dielectric layer induced by top electrode

In various TEM bright field images, induced strain to the crystalline $CaTiO_3$ from top electrodes can be identified. This is expressed by a brightness change for an otherwise homogeneously crystallized layer [148, 149, 150, 151]. Figures 6.2 a) and 6.2 b) show TEM images of MIM capacitor stacks, where lattice strain from the top electrode is visible. The crystalline $CaTiO_3$ reveals its typical (220) lattice planes with a plane distance of $1.36\,Å$. The parallel planes evolve at the bottom electrode up to the top electrode, indicating a single crystallite. Additional periodic changes in intensity appear when reaching the top electrode. This periodicity has been emphasized in the insets of both pictures. The continuation of the d_{220} lattice planes (besides appearance of brightness changes) indicate a crystal rotation around the [220] lattice vector. The crystal rotation seems to be necessary to adapt to the top electrode lattice. The adaption appears for the two samples with different crystallization process (high temperature deposition vs. PDA), which indicates an induced strain independent of the used process. The values of periodicity are larger than one unit cell of cubic $CaTiO_3$. Nevertheless the length scales can be interpreted as multiples for the d_{111} plane ($2 \times d_{111}$) for the sample with TiN top electrode and multiples for the d_{110} plane ($4 \times d_{110}$) (Table 6.1) for the sample with Ru top electrode. There is a potential correlation between the lattice of both the top electrode and $CaTiO_3$. The brightness maxima can be assigned as the least common

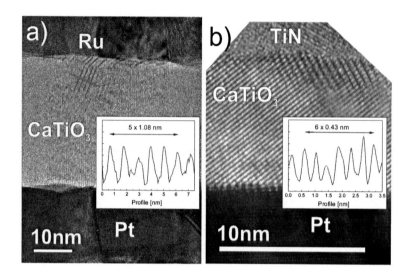

Figure 6.2: TEM images of MIM capacitor stacks showing strain in dielectric layer induced by top electrode (red lines). a) Stack with CaTiO₃ between Pt and Ru electrodes deposited at 550°C (Ru electrode at 100°C) b) Stack with CaTiO₃ between Pt and TiN electrodes deposited at 300°C and annealed at 600°C. The change in intensity is the result of a twisted unit cell around the [110] crystal direction of the visible lattice plains. The strain results in periodic change of intensity (blue lines in insets), which is significantly larger than the CaTiO₃ lattice constant ($a_{CTO} = 3.84$Å).

Table 6.1: Lattice plane distances for various cubic $CaTiO_3$ lattice planes (lattice constant $a = 3.84$ Å).

	(100)	(110)	(111)	(220)	(222)	(210)
d (Å)	3.84	2.72	2.22	1.36	1.11	1.72

denominator of both lattices in the given crystal orientation. No lattice strain induced by the Pt bottom electrode indicate an already relaxed $CaTiO_3$ layer due to the matching lattice of both materials. No strain has also been observed in TEM for a stack without top electrode (e.g. later in Figure 6.5).

Periodic brightness changes can be caused by other imaging effects of the layer structure. A brightness change have been reported in Ref. [152], where it has been correlated to stacking faults due to local stoichiometry variations. This explanation is unlikely here, because the layers match the $CaTiO_3$ stoichiometry (see Section 5.3). Another possibility are Moiré pattern as interferences between the lattice planes of different $CaTiO_3$ crystallites lying above each other. This can be excluded here, the pattern should than appear at grain boundaries in the $CaTiO_3$ layer and should not be dependent of the choice of electrode material. There is a correlation between the periodic brightness change in $CaTiO_3$ and the lattice direction in the top electrode (visible in Figures 6.2 a) and b) and following), induced strain is therefore the most reasonable explanation.

6.3 Comparison of post deposition anneals and CaTiO₃ crystallization during growth

Post-deposition anneals (PDA) of $CaTiO_3$ capacitor stacks are possible with samples containing TiN as top electrode. Figures 6.3a) and 6.3b) show TEM micrographs of a $CaTiO_3$ capacitor, where an amorphous layer of $CaTiO_3$ (11 nm) was annealed in N_2 for 1 min at 600 °C in N_2 atmosphere. The $CaTiO_3$ layer is completely crystallized. The crystallinity signal of $CaTiO_3$ is intensified with a dark field TEM in Figure 6.3b). The $CaTiO_3$ crystallites are significantly larger in lateral dimension compared to the film thickness. Corresponding capacitance measurements of the MIM capacitors (Figure 6.4) reveal a $CaTiO_3$ permittivity of 25-30 for amorphous $CaTiO_3$ and 90 after PDA above 600 °C for the sample in Figure 6.3. A series of samples with different deposition temperatures is included in Figure 6.4, showing beside the amorphous and crystalline permittivity an intermediate permittivity of approximately 50 starting at 550 °C. This

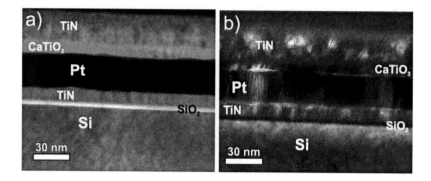

Figure 6.3: TEM images of CaTiO₃ capacitor stack annealed at 600°C. The crystallization of CaTiO₃ is visible in a) a bright field STEM as well as b) corresponding dark field STEM image. 10 nm CaTiO₃ has been deposited at 300°C and annealed at in N₂ atmosphere for 1 min. The CaTiO₃ layer is fully crystallized, some parts are visible as darker part in CaTiO₃ of Figure 6.3a) or as brighter part in Figure 6.3b).

has not been observed in any sample prepared with PDA. One sample of CaTiO₃ on Pt deposited at 600 °C without a top electrode is depicted in Figure 6.5. Only partial crystallization of the CaTiO₃ layer is visible. The sample is 5 nm thinner but has been deposited at a higher temperature compared to a previous sample in Figure 6.1. The layer thickness may be more relevant than the deposition temperature to achieve crystallinity of CaTiO₃. The relationship between layer thickness and crystallization temperature was investigated for amorphous Si/SiO₂ superlattices in Ref. [155] and modeled later in Ref. [156]. The model from Ref. [156] is shown in Equation (6.1) and includes an exponential progression with the layer thickness. The crystallization temperature T_{crys} is expressed as

$$T_{crys} = T_{ac} + (T_{melt} + T_{ac}) \exp^{-d/C} , \qquad (6.1)$$

with T_{melt} describing the melting temperature of the bulk crystal, T_{ac} is the crystallization temperature of a thick amorphous film, d is the thickness of the layer and C a constant. This dependence has been confirmed for ultrathin layers of Ge₂Sb₂Te₅ films (5-20 nm) in Ref. [157] and may be adopted to the different amount of crystallites in the CaTiO₃ layers investigated here. 35 nm CaTiO₃ layers deposited at 550°C on Pt show a higher degree of crystallization compared to 30 nm CaTiO₃ deposited on Pt at 600°C. This small variation in temperature indicates that the temperature dependence to form

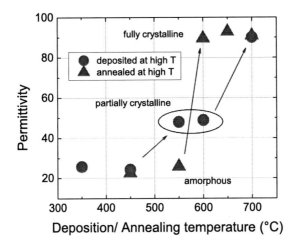

Figure 6.4: Permittivity of CaTiO₃ at different deposition/ annealing temperatures. The samples deposited at high temperatures (30 nm CaTiO₃ thickness) exhibit an intermediate k-value of approx. between 550-600°C T_{dep} (marked blue). At a T_{dep} of 700°C, the permittivity reaches a value of 95. Annealed CaTiO₃ layers show an immediate increase of permittivity to 90 above T_{an} larger than 550°C without a sign of partially crystalline CaTiO₃.

Figure 6.5: A CaTiO₃(30 nm)/Pt(42 nm)/TiN(3 nm) stack on top of a Si substrate shows only partial crystallization of CaTiO₃. A small interface at the top of CaTiO₃ between the preparation layer (Pt) indicates a reaction of CaTiO₃ with ambient gases (XPS of surface in Figure 5.7) or the formation of hydroxides [153, 154].

71

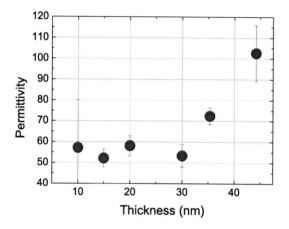

Figure 6.6: Thickness dependence of CaTiO₃ permittivity at 550 °C deposition temperature. The CaTiO₃ layers up to 30 nm exhibit a permittivity of approx. 55. The further increase in layer thickness results in an increase of permittivity to 104. The permittivity values have been extracted from a linear fit of the inverse capacitance over thickness by neglecting any interfacial capacitance.

crystallites is very susceptible in the region of T_{crys} of CaTiO₃. This susceptibility is more pronounced in capacitance measurements on a CaTiO₃ thickness series at a constant deposition temperature.

Figure 6.6 shows the thickness dependence of permittivity of CaTiO₃ deposited at 550 °C on a Pt electrode. The permittivity values have been extracted from a linear fit of the inverse capacitance over thickness by neglecting any interfacial capacitance. The thin layers reveal a permittivity of approx. 55. With increasing the layer thickness, the permittivity increases up to 104. The differences in permittivity with layer thickness is the result of partially crystallized CaTiO₃ at the used deposition temperature. To confirm this statement, the samples with intermediate permittivity are now further investigated in detailed TEM analyses.

6.4 Nanocrystals of $CaTiO_3$ in an amorphous matrix

Previously, $CaTiO_3$ within or without a MIM structure has been deposited at a high deposition temperature and/or sufficiently thick to achieve crystallinity. The reason for the inhibition of crystallization is due to the increasing contribution of the surface energy to the overall energy, when going to thinner samples [156]. It should be noticed that in Ref. [156] the partial crystallization of Si nanocrystals in an amorphous SiO_2 matrix occurred as a phase separation of two materials with different melting points. The physics behind the formation of $CaTiO_3$ nanocrystals may differ.

Reducing the deposition temperature and the thickness of $CaTiO_3$ does not simply lead to an amorphous dielectric. It shows an intermediate state of crystallization, where parts of the layers are amorphous and crystalline (Figures 6.5 and 6.7b)). Partial crystallization has also been observed by Strobl et al. [158] and Dunlop et al. [159]. It had been either done under phase separation or locally applying enough energy for crystallization with ion bombardment. These explanations to not apply to the observed partial crystallization in the $CaTiO_3$ layers. The temperature is applied equally throughout the layer and phase separation is unlikely. No change in band gap from previous LE-EELS measurements indicates an uniform composition for amorphous and crystalline $CaTiO_3$ (Figure 5.4). This hints towards an influence of the underlying Pt layer, which will be investigated consecutively.

For equal MIM capacitor stacks with different dielectric thickness, a deposition temperature of 550 °C leads to full crystallization (Figure 6.1) for a sample with thick oxide larger than 30 nm. For the thinner sample with equal deposition conditions, most parts of the $CaTiO_3$ layer are amorphous (Figure 6.7 a)), crystallization occurs only in some small areas (Figure 6.7 b)). Most of these crystallites show the same lateral size as the underlying crystallites in the Pt electrode. This hints towards a lattice correlation between the Pt electrode and the $CaTiO_3$ layer.

Previous experiments showed epitaxial growth of $SrTiO_3$ as a comparable perovskite on top of various Pt surfaces [160, 161, 162]. Nevertheless, $SrTiO_3$ has a lattice constant of 3.905 Å (eg. Ref. [163]), which has a better lattice match towards Pt with a lattice constant of 3.92 Å compared to a value of 3.84 Å for $CaTiO_3$. As literature shows, the two per cent difference between the Pt and $CaTiO_3$ lattice constants typically introduce enough lattice strain to the oxide layer to prevent any structural ordering [164]. The lattice strain and the crystallization inhibition due to the thin oxide layer (Equation (6.1)) may be the main reason for the amorphous part of the $CaTiO_3$ layer.

Obviously, this is not valid for the crystallites surrounded by the amorphous matrix in

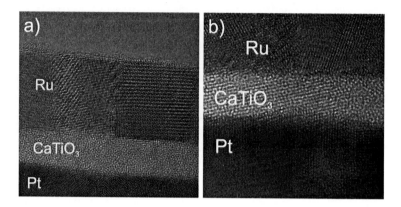

Figure 6.7: TEM micrographs from 5 nm CaTiO₃ layers show partial crystallinity. The sample has been deposited at 550°C on top of a Pt electrode. Most parts are amorphous (Figure 6.7a)). Some crystalline parts are visible with a lateral dimension matching the underlying Pt crystal (Figure 6.7b)).

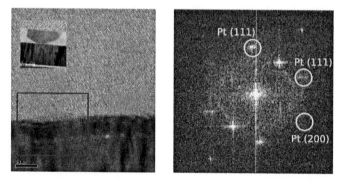

Figure 6.8: TEM image of bottom of inverse pyramidal shaped crystal on top of Pt. The corresponding reciprocal space from the rectangle show both reciprocal lattice points for Pt and CaTiO₃.

 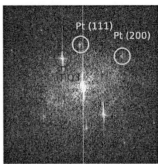

Figure 6.9: Another TEM image of a nanocrystallite on top of Pt. The reciprocal image shows an equal correlation of lattice point as in Figure 6.8.

Figure 6.7 b). The size of the nanocrystals correspond to the underlying Pt crystallites. An even more predicated sample has been shown in Figure 6.5, where the Pt crystals in the electrode work as seed crystals for the formation of inverse-pyramidally shaped $CaTiO_3$ crystals. TEM images at the interface in Figures 6.8 and 6.9 reveal, that all measured $CaTiO_3$ crystallites grow on a specific oriented Pt crystallite. The $CaTiO_3$ lattice planes show the same correlation for the planes of the Pt crystallite. The reciprocal space images in Figures 6.8 and 6.9 back that correlation.

For simple cubic unit cell materials, it is simple to identify the exact correlation between those two materials. It is rather complicated to identify this in the $Pt/CaTiO_3$ system, especially for the orthorhombic $CaTiO_3$ unit cell (twisted cubic structure, Figure 6.10). First it is necessary to identify the Pt and $CaTiO_3$ crystal view direction in TEM and then the angle of correlation for both crystallites. Second, the angle finally hints towards the Pt surface, which is responsible for the ordered growth of another specially oriented $CaTiO_3$ crystallite.

To begin with, the orthorhombic unit cell of $CaTiO_3$ is given in Figure 6.10 together with its pseudo-cubic unit cells. This figure shows that the orthorhombic unit cell consists out of two pseudo-cubic cells stacked in z-direction and rotated by $45°$(cubic $\{110\}$ direction). In Figures 6.8 and 6.9, the Pt view direction is the $\{100\}$ direction of a fcc crystal. The visible atomic positions of Pt show reciprocal lattice points in both figures (right FFT images). The $CaTiO_3$ direction is more complicated to identify, only two reciprocal lattice points are available. The $CaTiO_3$ crystal is minimally tilted in the TEM images, the remaining visible planes indicate a $\{220\}$ view direction. The different points of view for $CaTiO_3$ are simulated in Figure 6.11. The visible planes of $CaTiO_3$ for the assumption

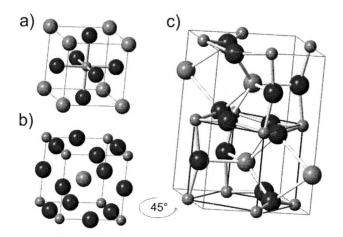

Figure 6.10: Simple cubic structure of CaTiO$_3$ (a = 3.84 Å) with a) Ti centered and b) Ca centered. c) Orthorhombic unit cell of CaTiO$_3$. The cubic cell is rotated by 45° around the z-axis. To include the tilted TiO$_6$ octahedrons, the unit cell size is double the size along z-axis in Pbmn structure.

of a cubic cell (a = 3.84 Å) can be calculated with

$$d_{hkl} = \frac{2\pi}{|\mathbf{g}_{hkl}|} \, ,$$

$$\vdots$$

$$d_{hkl}^2 = \frac{1}{\left(\frac{h}{a}\right)^2 + \left(\frac{k}{b}\right)^2 + \left(\frac{l}{c}\right)^2} \, ,$$

$$d_{hkl}^2 = \frac{a^2}{h^2 + k^2 + l^2} \tag{6.2}$$

with $a(,b,c)$ as the lattice constants, h,k,l as the integer values for reciprocal lattice points and $\mathbf{g}_{hkl}$ as the reciprocal lattice vector. d_{220} = 1.35 Å is the distance for the lattice planes visible in e.g. Figures 6.7b) and 6.8. In Table 6.2, all reciprocal lattice distances for the cubic and orthorhombic CaTiO$_3$ crystal are calculated[1] for different view directions. In most TEM images, one can identify the [220] plane of CaTiO$_3$, no single

---

[1]For cubic CaTiO$_3$, the lattice constant is a = 3.84 Å. For the orthorhombic structure, the lattice constants from Ref. [4] have been taken.

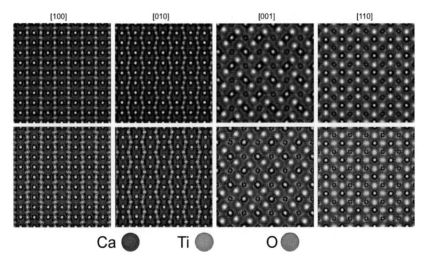

Figure 6.11: Simulation of TEM pattern for CaTiO$_3$ with different zone axes. As unit cell, the orthorhombic Pbmn crystal structure was taken. It is shown in Figure 6.10, that the orthorhombic unit cell is rotated about 45° around the z-axis compared to the standard simplified cubic structure. Therefore, the orthorhombic [100] and [010] zone axes correspond to the [110] zone axis of the cubic structure. Equal distances for the main (quadratic) features(/atoms) can be viewed from the [001] and [110] orthorhombic zone axes.

Table 6.2: Calculation of reciprocal lattice vectors from different view directions. It can be seen, that the distances in reciprocal space reappear for both the cubic and orthorhombic $CaTiO_3$ crystal structure or are multiples of it from different view directions. The different lengths in z-direction for the orthorhombic structure are the result of the rotated unit cell (see Figure 6.10).

View direction	Reciprocal lattice vectors of $CaTiO_3$ in Å^{-1}							
	{220} orth.	{110} cub.	{111} orth.	{111} cub.	{200} orth.	{200} cub.	{112} orth.	{112} cub.
[100]	0.452	0.368	-	-	0.369(020) 0.261(002)	0.521	-	-
[110]	0.261(110)	0.368	0.584(222)	**0.451**	0.261(002)	0.521	0.370	0.638
[111]	0.522(220) 0.452(022)	0.368	-	-	-	-	0.370(112) 0.434(211)	0.638

atoms (atom columns) are visible. This is explicitly shown in reciprocal space, where reciprocal lattice points in only one direction with $|\mathbf{g}_{111}|$ of 0.451 Å^{-1} can be extracted. Only a few images show more $CaTiO_3$ lattice points in reciprocal space, especially at the boundary of the isolated crystals. This hints, besides a crystal rotation perpendicular to the [220] plane, towards an off-center position of Ti atoms [34] or even a screw dislocation in $CaTiO_3$ [165].

Pt on the other hand typically shows various lattice points, from which the view direction and single planes can be identified. As conclusion, the oriented growth of $CaTiO_3$ on Pt is not achieved like $SrTiO_3$ on Pt along the {100} lattice planes as in Ref. [160], but a small shift in angle between both unit cells is persistent. Finally, the TEM images conclude, that only a Pt {111} interface leads to an oriented growth of a (111) $CaTiO_3$ crystal. For other oriented Pt surfaces no epitaxial relation could be identified in all experiments.

6.5 Discussion of preferential growth

Lattice parameters with surface energies of both Pt and $CaTiO_3$ are playing an important role for the ordered growth. The Pt {110} surface tends to form {111} facets at higher temperatures [166] with increasing roughness compared to the {100} and {111} surfaces [167]. On a surface heated up to 600°C, the fraction of Pt {110} facets is reduced compared to other facets.

So far, little is known about epitaxial behavior of $CaTiO_3$. Epitaxial recrystallization experiments revealed an activation energy of 3.18 eV for [010] and 3.89 eV for [100] surface direction of $CaTiO_3$ [168]. Various simulations were performed using ab-initio methods to investigate the surface energy of $CaTiO_3$ and to gain knowledge of surface relaxation for different crystal directions. In Table 6.3, literature values of these simulations for surface energies per unit cell of $CaTiO_3$ are given. The surface energy E_{surf} is simply the sum of the cleavage energy E_{cleav} of the crystal and the relaxation energy E_{relax} [21]:

$$E_{surf} = E_{relax} + E_{cleav} \qquad (6.3)$$

The calculated results point to the surface energy of a $CaTiO_3$, which is highest for the {110} and {111} surfaces. All ABO_3 perovskites ($CaTiO_3$, $SrTiO_3$, $BaTiO_3$, $PbTiO_3$, $SrZrO_3$, $BaZrO_3$, ...) show that the (001) surface always exhibits the lowest surface energy. The main physical reason for the lowest surface energy for (100) is that the {100} surface consists out of neutral planes (CaO with Ca=+2e ionic charge, O=−2e) and TiO_2 (Ti=+4e and O=−2e) (see planes in Figure 6.11). These calculations reveal that the (111) surfaces for all ABO_3 are the most unstable ones and have the largest surface energy [21]. The $CaTiO_3$ surfaces in (110) and (111) directions are both polar (no neutral planes), and therefore react differently in comparison to the (100) surface. Results from literature [161, 169] evaluated that metals adsorb more strongly on polar surfaces than on non-polar surfaces.

The Figures 6.8 and 6.9 show the (220) $CaTiO_3$ crystallite lattice points in a specific orientation to the Pt surface. In both TEM images, the Pt (111) facet is always the foundation of the formation of these $CaTiO_3$ crystallites. No growth of $CaTiO_3$ crystallites on {110} or {100} Pt surfaces was observed. The conclusion is that, beside the good lattice match between (111) Pt and (111) $CaTiO_3$, that the {111} surface of $CaTiO_3$ is the most unstable surface and forms the strongest bonds towards the Pt surface according to the $SrTiO_3$/Pt system in Ref. [169]. In that case, the formation energy of $CaTiO_3$ crystallites is significantly reduced compared to the other facets. When going to higher deposition temperatures or thicker samples (Figure 6.1), the preferential growth along the {111} surface can not be distinguished in the entire crystallized $CaTiO_3$ layer.

For Pt deposition on $SrTiO_3$ shown in Refs. [161, 169], with 1 - 3 monolayers (ML) of epitaxially grown Pt, the work of separation (the energy required to separate both crystals) is largest for the (110) surface of $SrTiO_3$. The energy depends on the Pt-O bindings at the interface, an O-terminated $SrTiO_3$ binds strongest to Pt. This can be the case for the preferred growth of $CaTiO_3$ on Pt(110) with a O-terminated $CaTiO_3$ surface (Figure 6.11). DFT calculations from Ref. [169] finally show, that for (620) O-terminated surfaces of $SrTiO_3$, the work of separation is highest. In [174], the small

Table 6.3: Simulated surface energies E_{surf} of cleaved and relaxed $CaTiO_3$ surfaces for various surface terminations. The overall surface energy is lowest for the {100} plane direction. {110} surface shows an energy level close to the one for {111} surfaces.

E_{surf}	Eglitis [20, 170] (eV/cell)	Zhang [171] (eV/cell)	Wang [172] (eV/cell)	Liu [173] (eV/cell)
(100)	1.13 TiO_2	1.021 TiO_2	0.95 TiO_2	-
	0.94 CaO	0.824 CaO	0.76 CaO	-
(011)	3.13 TiO	2.180 TiO	-	
	1.91 Ca	1.671 Ca	-	
	1.86 O	3.074 B-type O	-	
(111)	5.86 CaO_3	-	-	3.993 CaO_3
	4.18 Ti	-	-	3.411 Ti
Method	DFT-B3LYP B3PW	DFT + GGA - BFGS	GGA - BFGS	LDA

derivated (621) surface has been used for epitaxial deposition of (621) Pt. The possibility to resolve these high Miller index surfaces by TEM can not be excluded, but are unlikely under the given experimental conditions.

6.6 Conclusion

Four $CaTiO_3$ layers within a MIM structure or without a capping layer were selected for detailed investigation with TEM under targeted deposition conditions. Two samples with 5 nm and 35 nm thickness have been deposited at 550 °C on top of a Pt electrode and received a Ru top electrode deposited at 100 °C. One 30 nm $CaTiO_3$ layer has been deposited at 600 °C on a thick and smooth Pt electrode without a capping layer. One 10 nm thin $CaTiO_3$ layer on top of Pt was deposited at 300 °C with a top electrode also deposited at 300 °C. This sample received a PDA at 600 °C in N_2 atmosphere for 1 min. With these samples, an overall statement for the other comparable experiments is possible.

There is a preferential growth of $CaTiO_3$ crystals on top of a Pt (111) surface facet resulting in isolated crystals in an otherwise amorphous matrix. The influence of the isolated crystals on electrical properties was addressed in Ref. [175]. Going to higher deposition/annealing temperatures and layer thicknesses, the layer becomes fully crystallized and the proportion of preferential $CaTiO_3$ orientation is masked by unordered crystallites.

The use of electrodes with a major deviation in lattice constants to $CaTiO_3$ result in strain induced to the dielectric. This strain influences the crystallization temperature. No lattice strain was seen for Pt as electrode material with its good lattice match to $CaTiO_3$. The result is that the $CaTiO_3$ crystallization temperature on Pt is lower compared to samples with lattice mismatching surfaces.

7 Electrical characterization of calcium titanate in MIM capacitor stacks

7.1 The relation between physical and electrical properties of CaTiO$_3$ MIM capacitors

In the previous Chapters 4, 5 and 6, a detailed investigation of physical properties of CaTiO$_3$ and the electrodes was presented. Their possible influence on the electrical properties of CaTiO$_3$ capacitors has already been addressed briefly, but shall be described here in more detail.

Rough electrodes will increase the leakage current especially for ultrathin capacitors due to electric field enhancements at the protrusion of the electrodes (e.g. in Ref. [69]). The roughness locally increases the electrical stress to the CaTiO$_3$ and leads to low breakdown voltages [176, 177]. Therefore the reduction of the surface roughness of these bottom electrodes is expected to decrease the leakage current significantly and was one main objective of suitability tests of bottom electrode investigation. The control of roughness is even more important for high temperature processes. Those are required to increase the degree of crystallinity of CaTiO$_3$ during deposition or through post-deposition annealing (PDA).

Capacitance measurements are not directly influenced by the electrode roughness. The extracted capacitance can be influenced by the different defect concentrations induced by the electrode roughness. Measurements of capacitance at a bias voltage causing high leakage currents result in dissipation factors above the chosen limit of 0.1 for the parallel model. The voltage ranges have than been reduced. If the capacitance voltage measurements show dissipation factors above 0.1 for every measurement point, the results have been disregarded (see Experimental Section 3.3).

Contaminations of the deposited CaTiO$_3$ layers can form trap states in the band gap of CaTiO$_3$ and reduce the effective barrier height by a substantial level. This will increase the leakage current of the capacitor significantly. From XPS measurements in Chapter

5, no evident contamination has been found in the layer, which could influence the following leakage current measurements. Nevertheless, a large oxygen deficiency is revealed by XPS (Section 5.2), which would increase the trap density and may form a trap band below the conduction band [74, 178, 179]. On the other hand, ellipsometry data as well as EELS scans showed no reduction of band gap due to trap states (Section 5.1). While roughness, defects and vacancies are of major importance for the leakage current, the nature of crystallinity of $CaTiO_3$ plays a significant role to capacitance and permittivity. As indicated in Chapters 5 and 6, the crystallinity of $CaTiO_3$ is achieved at approx. 600 °C in dependence on the electrode composition. Some layers showed a mixed composition of amorphous and crystalline $CaTiO_3$, which is expected to result in lower effective permittivities of the capacitors. Additionally, the TEM images showed no evidence of an interfacial layer formation between the $CaTiO_3$ and the top and bottom electrodes. This is important for the overall capacitance of the capacitors, because without low-k interfacial layers like TiON (Ref. [113]), the dielectric permittivity is not reduced. With these facts, thickness scaling of capacitors with fully crystallized $CaTiO_3$ is expected to alter capacitance linearly, as in an ideal capacitor.

In this chapter, the results of electrical characterization start with a detailed investigation of the capacitance density and permittivity of $CaTiO_3$ capacitors. The current voltage dependence on various electrodes and at different temperatures is described afterwards.

7.2 Capacitance measurements of $CaTiO_3$ capacitors

The investigation of the capacitance of $CaTiO_3$ capacitors was split in four parts:

1. At first, the capacitance measurements were done on different electrodes with various thicknesses.

2. Additionally, the influence of the applied electric field on the permittivity of $CaTiO_3$ has been investigated in detail.

3. Temperature dependent $C(V)$ investigation shall further reveal the capacitors behavior to $CaTiO_3$ permittivity and their field dependent parameters.

4. Frequency dependent $C(V)$ measurements complete the picture of $CaTiO_3$ in frequency ranges from 40 Hz to values above 1 MHz.

For capacitance measurements, the standard setup as described in Experimental Section 3.3 has been used to investigate the permittivity of $CaTiO_3$.

7.2.1 Capacitance dependence on thickness

In a capacitor stack, the capacitance density is inversely proportional to the thickness according to the capacitance equation of a plate capacitor in Equation (7.1)

$$C = k\,\varepsilon_0\frac{A}{d} \tag{7.1}$$

with as the dielectric permittivity k, ε_0 is the vacuum permittivity, A is the area of the capacitor and d is the thickness of the dielectric. This equation allows a simple validation of experimental results, especially of the correct thickness for the ultrathin dielectric layers. Because the measured capacitance comprises both a 'bulk' capacitance and an interface capacitance (e.g. death layers), the capacitance may not follow to the reciprocal thickness dependence in Equation (7.1). Therefore, a the serial arrangement of a film C_{film} and interface capacitance C_{int} is presented in Equation (7.2).

$$\frac{1}{C} = \frac{1}{C_{int}} + \frac{1}{C_{bulk}} = \frac{d_{int}}{A\,\varepsilon_0 k_{int}} + \frac{d_{film}}{A\,\varepsilon_0 k_{film}} \tag{7.2}$$

According to Equation (7.2), the inverse capacitance increases linearly with the thickness of the film (for $d_{int}, k_{int} = const$). Any interfacial layer between CaTiO₃ and an electrode would result in a distinguishable vertical shift.

In Figure 7.1, capacitance data from amorphous CaTiO₃ capacitors with thicknesses up to 90 nm is depicted in an inverse capacitance versus thickness plot. Following Figure 7.1, the linear fit traverses the point of origin. The extracted k-value for amorphous CaTiO₃ is 26. No vertical shift is visible, a contribution of an interfacial layer with relevant C_{int} can therefore be neglected and $C_{film} \approx C$.

A CaTiO₃ thickness series of capacitors with (partially) crystalline CaTiO₃ has been discussed in detail in the previous chapter in Figure 6.6. There, the CaTiO₃ layers exhibit permittivities between 50 and 90. The difference in permittivities had been assigned to a partial crystallization of CaTiO₃. These permittivities are significantly larger compared to the permittivity of ZrO₂ ($k \approx 24 - 40$ [2]) currently integrated in DRAM structures. Most of the fully crystalline CaTiO₃ samples exhibited high leakage currents (see later in the conduction part of this chapter) and prevented a reasonable extraction of capacitance densities (dissipation factors $D > 0.1$). With process optimization and the use of temperature stable bottom electrodes (Chapter 4), samples with crystalline CaTiO₃ have been successfully prepared and characterized. As example, CaTiO₃ exhibits a permittivity of 105 with a leakage current of $6.8 \times 10^{-9}\,A/cm^2$ on a carbon electrode. Other samples with crystalline CaTiO₃ are discussed later (e.g. Figure 7.6 or Figure 7.8).

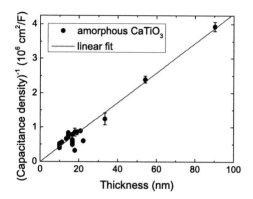

Figure 7.1: Inverse capacitance density versus thickness of amorphous CaTiO$_3$ capacitor stacks. The linear fit (red line) of the measurement data has an intercept close to zero, thus no interfacial layer is present.

7.2.2 Non-linear voltage dependence of capacitance

A voltage sweep has been applied to the capacitor to evaluate any field dependent change in capacitance. Perovskites and other oxides typically exhibit an electric field dependent change of permittivity [64, 180, 181]. In this work, capacitance voltage ($C(V)$) measurements of CaTiO$_3$ capacitors show the non-linearity for CaTiO$_3$. This non-linearity can be described with the polynomial in Equation (7.3),

$$C(V) = \sum_i^n a_i V^i = a_0 + a_1 V + a_2 V^2 + \ldots + a_n V^n \approx C_0(1 + \beta V + \alpha V^2) \quad (7.3)$$

with a_i, β, α as constant coefficients and C_0 as the extremal capacitance. This is shown in Figure 7.2 for the 2nd to 4th order fitting curves including the residuals (Figure 7.2b)). The error of fitting is reduced with any higher order approximation. With fitting a N order polynomial, typically a larger order residual remains. In literature, higher order terms above $N = 2$ have been neglected due to simplicity reasons. The low errors in Figure 7.2b) validate the approximation for CaTiO$_3$. The 2nd order approximation in Equation (2.5) has further been used to extract the voltage dependent constants of capacitance (VCC), namely α for the second order and β for the first order voltage dependence of capacitance.

As described in Section 2.4, non-linearity in $C(V)$ measurements can arise for various

(a)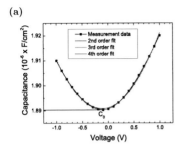

(b)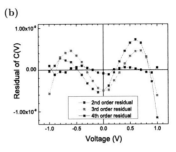

Figure 7.2: Capacitance vs. voltage for a 10 nm amorphous Pt/CaTiO₃/Ru capacitor. a) Fitting of measured $C(V)$ curve with a 2nd to 4th order polynomial. The minimum is shifted to negative voltages. An increasing order result in a better fit of the measured data. b) Residual as a function of voltage for the 2nd to 4th order polynomial fit. The error is significantly reduced for higher order polynomials.

reasons, e.g. ionic or orientation polarization, electrostriction of the dielectric or the oxygen affinity of the electrode. The results obtained for CaTiO₃ in this work shall be compared to the results of related and competing materials studied elsewhere. This includes the comparison of own results with available theoretical models.

In order to extract the coefficients of capacitance α and β in Equation (7.3) independent from capacitance density, the $C(V)$ results have been normalized to ΔC according to Equation (7.4)

$$\Delta C = \beta V + \alpha V^2 = \frac{C(V) - C_0}{C_0} . \tag{7.4}$$

The extracted α and β are summarized for amorphous layers with various electrode stacks and deposition/ annealing temperatures in Figure 7.3a) and 7.3b) respectively. Figure 7.3a) shows a reciprocal behavior of the used MIM capacitor stack. There is no deviation visible for different electrode arrangements. With the different interfaces for various metal/oxide interfaces, it can be assumed that α is an intrinsic property of CaTiO₃ and not an interface dependent property. The small increase of α when reducing the deposition/annealing temperature in Figure 7.4a) hints towards an α dependence on the layer quality. A possible explanation is that defect densities and Ar incorporations originating from the sputter process are reduced, when increasing the deposition temperature of CaTiO₃. While no electrode dependent properties can be recognized for α, the opposite is valid for β. As shown in Figure 7.3b), different electrodes have a different influence on the value of β. Beside the decreasing displacement of the $C(V)$ curve along

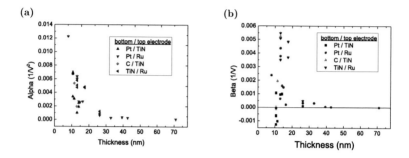

Figure 7.3: (a) α and (b) β versus CaTiO₃ thickness of amorphous capacitors with different electrodes. The alpha factor increases with reducing the thickness independent of the used capacitor stack. β shows a huge spreading for thin layers of CaTiO₃. Layers containing Pt show a trend of β close or below zero.

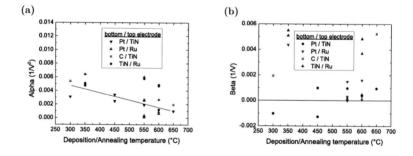

Figure 7.4: (a) α and (b) β versus deposition/annealing temperature in various electrode stacks. α increases linearly with decreasing deposition temperature. For the thinnest layers of CaTiO₃ and temperatures close to the crystallization temperature of CaTiO₃, the value of α is distinctively increased. In Figure 7.4 β shows a huge spreading and no trend for different deposition temperatures.

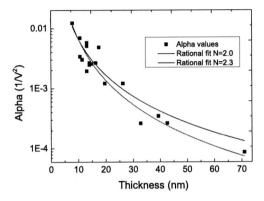

Figure 7.5: Logarithmic α dependence vs. thickness. The data is fitted with a rational as well as an exponential fit. The best agreement to measurement data is achieved with $\alpha \propto x^{-2.3}$.

the abscissa axis with increasing $CaTiO_3$ thickness, a clear reduction of β is depicted when Pt as bottom electrode material is involved. A significant shift in the value towards negative voltages ($\beta > 0$) is identified for all other electrode arrangements. For Pt, the employed electrode with the largest work function, the difference towards the work functions of TiN or Ru are comparatively large (approx. $1.0\,eV$). Any dependence of β from the difference in work function of both bottom and top electrodes should result in a larger shift for capacitor stacks containing Pt and an insignificant shift for electrodes with comparable work function (TiN, Ru, C). Nevertheless, the results show a completely reversed behavior. The observed influence of Pt hints towards the good lattice match of the Pt bottom electrode towards $CaTiO_3$. The better interface formation on the bottom electrode for the amorphous $CaTiO_3$ layer seems to result in the reduction of β.

The capacitor stacks containing Pt as bottom electrode always include a 'misfitting' top electrode. The reduced value of β for Pt and large value for the other electrodes concludes that only the interface between the bottom electrode and the deposited $CaTiO_3$ is important for β. No influence from deposition or annealing temperatures on β can be derived from Figure 7.4b). The values are scattered without revealing a tendency.

While derivations of β have been disregarded in literature, various publications deal with explanations for α (Section 2.3). Experiments with amorphous dielectrics in Ref. [57] show a reciprocal quadratic dependence of α with thickness. To compare the results

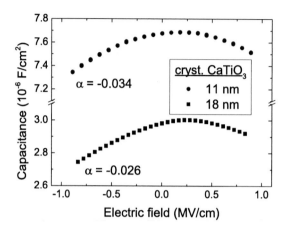

Figure 7.6: Field dependence of capacitance for two capacitors containing crystalline CaTiO₃. In contrast to the amorphous or semicrystalline samples, α is negative and approx. ten times larger compared to amorphous CaTiO₃ layers with comparable thickness.

with CaTiO₃, some fits to α from Figure 7.3a) are depicted in Figure 7.5.

$$\alpha \propto \frac{1}{\left(\frac{x}{nm}\right)^N} \qquad (7.5)$$

The rational fit with $N = 2$ (Equation (7.5)) shows a good agreement to the measured values of α. The fit slightly overestimates α for thicker oxide layers. The fit with $N = 2.3$ shows the lowest error.

According to Ref. [57], the α dependence on thickness is caused by electrostriction and reveals a quadratic thickness dependence. There the thickness dependence has been approximated by neglecting higher order thickness dependencies. In the case of CaTiO₃, the used quadratic polynomial is only an approximation of a higher order polynomial, which has been addressed previously in Figure 7.2.

To complete the picture of α for crystalline CaTiO₃, two completely crystalline layers of CaTiO₃ in Figure 7.6 with 11 nm and 18 nm thickness are presented. The $C(V)$ measurements reveal a negative α different to the amorphous and semicrystalline samples in Figure 7.3. Both capacitors with TiN/Pt bottom and TiN top electrode have been annealed at 650 °C. The absolute value of α is with |-0.034| 1/V² for the 11 nm oxide approximately ten times larger compared to amorphous capacitors with similar thickness. Several reasons for the different curvatures of amorphous and crystalline CaTiO₃ layers have been presented in Section 2.3. The change from positive to negative α values may be the result of arising long-range potentials in crystalline CaTiO₃. The capacitance drop at an applied voltage is a known behavior for Ba$_x$Sr$_{1-x}$TiO₃ [56, 182] or SrTiO₃ [183] capacitors. It is significantly lower for CaTiO₃. The capacitance for fully crystalline CaTiO₃ capacitors is nearly unchanged with an applied electric field.

The extracted α for CaTiO₃ with approx. 34.000 ppm/V² for 11 nm crystalline CaTiO₃ is significantly larger than the limit for RF applications specified in the ITRS road map [184]. Various approaches have been investigated to reduce the oxygen vacancy concentration and with it the α value in literature. The anneal of SrTiO₃ with N⁺ helped to reduce α by a factor of 4 [185]. Similar results were obtained, when SrTiO₃ samples were annealed in oxygen atmosphere (factor 3-4 better compared to a nitrogen anneal) [186]. Another approach was the introduction of bilayers with additional ZrO₂ (positive α) to counteract the negative α of SrTiO₃ [187]. Because no efforts have been made to reduce α for CaTiO₃ in this work, all presented possibilities to reduce α can be applied to CaTiO₃ capacitors to reach the requirements of the ITRS road map.

7.2.3 Temperature dependence of capacitance

The temperature dependence of capacitance/ permittivity of perovskites has been investigated since 1950 [28]. The electrical behavior of the materials resulted in the development of crystallographic models, which are still valid nowadays. Barrets formula (Equation (2.1) in Section 2.1.3) addresses this temperature dependence and is derived from Slaters crystallographic model [188] of vibrating Ti atoms in a BaTiO₃ crystal with fixed Ba and O atom positions above the Curie temperature. With increasing temperature, the vibration forces of Ti atoms are reduced (broadened), which result in a decrease of permittivity. This would be the expected typical behavior of a dielectric.

With the purpose of integrating perovskites in devices with nanoscopic dimensions, new properties, like defect states, interfaces and oxygen vacancies arise and play a significant role in the dielectric behavior. These defects may be activated/suppressed in $C(V)$ measurements with changing temperature and frequency (frequency dispersion described later). In literature, both an increase or decrease of capacitance with increasing temperature with respect to the dielectric materials, deposition technique/temperature, stoichiometry and defects is described. An example for a decrease of permittivity with increasing temperature in polycrystalline CaTiO₃ and other perovskites is shown in Refs.

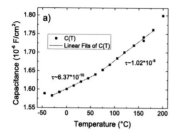
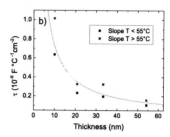

Figure 7.7: Temperature dependence of capacitance for amorphous CaTiO₃ capacitors. a) A linear increase of capacitance with temperature is observed for 10 nm as well as larger thicknesses. b) The slope τ from a) and capacitors with different oxide thicknesses is shown. A reciprocal dependence of the slope with the thickness can be assumed (grey line).

[15, 189]. Here, Figure 7.7a) shows the temperature dependence of capacitance $((C(T))$ of amorphous CaTiO₃. Because CaTiO₃ exhibits an increase of capacitance with increasing temperature, only literature focusing on this behavior is summarized. The temperature coefficient of capacitance (TCC) τ is the slope of the linear fit for different capacities with temperature T according to

$$\frac{C(T)}{C_0} = \tau T. \tag{7.6}$$

As seen in both plots of Figure 7.7, $C(T)$ measurements of amorphous CaTiO₃ reveal a linear increase of capacitance. The slope τ increases at $\approx 55\,°C$ from $6 \times 10^{-10}\ Fcm^{-2}\ °C^{-1}$ to $1 \times 10^{-9}\ Fcm^{-2}\ °C^{-1}$ for 10 nm oxide thickness. Other thicknesses show a similar change in τ at the same temperature (Figure 7.7b)). The positive values for τ are contrary to incipient ferroelectrics above the Curie temperature according to Barret's formula [28]. Nevertheless Blonkowski et al. [64] revealed a linear increase of capacitance for amorphous SrTiO₃. Some literature suggests a trap related reason for SrTiO₃ capacitors, because they observed a reduction of τ with annealing the samples in N⁺ plasma [185]. In Ref. [59], a microscopic model for the temperature dependence of $C(V)$ has been developed. The increase in ionic or dipolar polarization appears due to the interaction with the nearest neighbors. The assumption is there that the permanent moment of the molecules increases linearly with temperature. This results in an temperature increase of capacitance. This model had been applied to amorphous Al₂O₃ capacitors [58].

As result, the capacitance of amorphous CaTiO₃ increases with increasing the tempera-

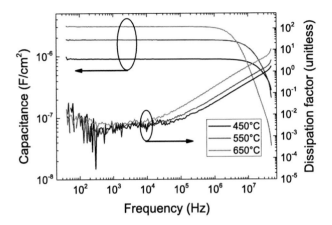

Figure 7.8: Impedance spectroscopy measurements at 0 V of 18 nm CaTiO₃ capacitors at different annealing temperatures. The increasing annealing temperatures result in increasing capacitance densities or k-values due to change from amorphous to crystalline layer structure. The measured capacitance reveals a small decrease up to 1 MHz for all samples.

ture. Comparable effects have been observed for amorphous SrTiO₃ capacitors and other oxides in literature. The assignment of the temperature increase to changes of local polarization effects hints towards an intrinsic property of CaTiO₃.

7.2.4 Frequency dependence of CaTiO₃ capacitance

Measuring the frequency dependence of capacitors reveal influences of defect contributions in the dielectric, like oxygen vacancies, contributing to the measured capacitance and the dielectric loss [82, 190]. It further allows statements about radio frequency (RF) performance and processing conditions of the deposited films [191, 192]. The measurement of capacitance in the frequency range between 40 Hz and up to 110 Mhz at 0 V allows the RF characterization of CaTiO₃MIM capacitors. The measurement was done with an impedance analyzer (HP LCR 4294A). The results for a sample with constant thickness and different annealing temperatures are shown in Figure 7.8. Three parts of one amorphous sample have been annealed at different temperatures (450°C, 550°C and 650°C). The different annealing temperatures induce different degrees of crystallinity from the amorphous to full crystalline $CaTiO_3$ layer in the capacitor stack. The different degrees of crystallinity result in different capacitance and therefore different k-values at a certain frequency. The frequency dependence for all samples is similar and shows a small decrease in capacitance for increasing frequencies up to 1 MHz. A large decrease of capacitance with an increase in the dissipation factor above 1 Mhz is due to the increasing contribution of the serial resistance from the used setup and is disregarded in the following discussion.

The experimental data allows the comparison to the frequency dependence of an ideal capacitor.

$$X_C = \frac{1}{2\pi f C} = \frac{1}{\omega C} \tag{7.7}$$

with X_C as the capacitive reactance, f as the frequency and C as the real capacitance. The small decrease in capacitance with increasing frequency in Figure 7.8 indicate that the frequency dependence of a $CaTiO_3$ capacitor does not fit to the one of an ideal capacitor. To specify the difference, a simple derivation from the ideal linear dependence in Equation (7.7) can be used to fit the results of Figure 7.8 with low error in Equation (7.8).

$$X_C = \frac{1}{(i\omega C)^\phi} \text{ with } (-1 \le \phi \le 1) \tag{7.8}$$

This model uses the principle of a constant phase element which produces an impedance having a constant phase angle $\theta = (\pi/2)^\phi$ in the complex plane [193]. For fitting, the program LEVM[1] has been used. As result for the $CaTiO_3$ capacitors from Figure 7.8, all the angles were $\theta = 89°$. The values are constant for all different states of crystallinity, indicating an intrinsic property of $CaTiO_3$. The difference to 90° for a perfect capacitor [81] is negligible.

---

[1]http://www.jrossmacdonald.com/levminfo.html

7.3 Leakage currents in $CaTiO_3$ capacitors

The investigation of leakage currents is fundamental to understand the performance of devices based on thin film capacitors, as described in detail in the Fundamentals (Section 2.4) and in the Preparation and Methods Chapter 3.1. The main goal of capacitor research, especially for metal-insulator-metal capacitors aiming DRAM applications, is the reduction of leakage currents with the optimal compromise between oxide layer thickness (and therefore capacitance density) and the leakage current limit at a chosen operation voltage (typically 1 V). The optimization of these capacitors with the highest achievable capacitance density at target leakage currents below $1 \times 10^{-7}\,A/cm^2$ at 1 V (ITRS road map [184]) are an important scope of this work and will be addressed here. The investigation of leakage currents in metal-insulator-metal capacitors made of $CaTiO_3$ is presented in the following order:

1. The influence of various electrodes on the conduction behavior of $CaTiO_3$ is investigated.

2. The thickness variation dependence of $CaTiO_3$ in the capacitor stack will be observed.

3. Temperature dependent $I(V)$ measurements are done to analyze the leakage mechanisms with respect to their temperature dependence. To improve the leakage currents of $CaTiO_3$ capacitors, an investigation of the conduction mechanism of $CaTiO_3$ capacitors is necessary.

4. The influence of semicrystalline $CaTiO_3$ (Chapter 6) on the conduction behavior in the capacitor is shown in detail.

7.3.1 The influence of different electrodes

The physical properties of electrodes, like work function and surface roughness, are essential for leakage current characteristics of any capacitor. High band gap oxides, like SiO_2 and Al_2O_3 do not depend on the work function of the used electrode material, because the conduction and valence band offset between the electrode and the oxide is sufficiently high to exclude any thermionic emission of charge carriers as an origin for leakage current [194]. In contrast, high-k materials with their comparatively low band gaps [47] need suitable electrodes with sufficiently high work function. Titanium nitride is a commonly used electrode material with a work function of 4.6-4.8 eV (see Section 2.2) for state-of-the-art capacitor applications with ZrO_2 or HfO_2 as high-k oxides [2].

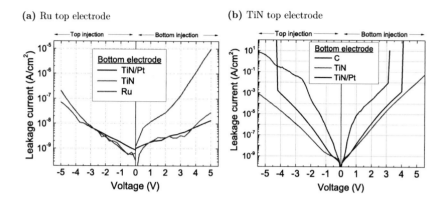

(a) Ru top electrode

(b) TiN top electrode

Figure 7.9: Leakage current as a function of voltage for different bottom and top electrode materials. a) Amorphous CaTiO$_3$ layers (30 nm thickness) have been deposited on top of comparatively rough electrodes with a Ru top electrode. b) Samples with TiN top electrode and amorphous CaTiO$_3$ (14.5 nm thickness) have been deposited on temperature stable electrodes including carbon.

To prepare capacitors with materials exhibiting a permittivity above 50, electrode materials with even higher work functions, e.g. noble metals, may be necessary to ensure a sufficient band offset above 1 eV [47]. The high costs for these materials, besides integration factors, has hindered their use in device applications. The leakage behavior of CaTiO$_3$ with its moderate band gap of 4.0 eV and the high permittivity of up to 186 in bulk crystalline material promises to be a good compromise and will be discussed in the following on different electrodes.

Results for leakage currents of amorphous CaTiO$_3$ on various bottom electrodes are shown in Figure 7.9. In Figure 7.9 a), leakage currents for capacitors with a 30 nm thick CaTiO$_3$ layer on top of TiN, Ru and Pt bottom electrodes have been measured. A Ru top electrode was used for all samples. Figure 7.9 b) shows leakage currents in capacitors with 14.5 nm CaTiO$_3$ thickness deposited on temperature stable C, TiN and Pt bottom electrodes with a TiN top electrode. All electrode substrates have been prepared prior to deposition of CaTiO$_3$ as described in Chapter 4. In Figure 7.9 a), the sample with TiN bottom electrode exhibits the highest leakage current compared to Ru and Pt electrodes for positive voltages applied to the top electrode. Both high work function materials Ru and Pt show comparable leakage currents, thus the barrier height towards CaTiO$_3$ is sufficient to suppress thermionic emission. Therefore bulk dependent defects are responsible for the main part of conduction in these samples. TiN and Ru exhibit a comparable work function (Section 2.2), the different leakage currents may be a sign of surface oxidation during the sputter deposition process of CaTiO$_3$ [195, 196]. This would be beneficial for the Ru surface (Ru transforms to RuO$_2$) and adverse for the TiN surface (TiN transforms to TiON$_x$)^2. However, other effects may also be responsible for the increase in leakage current, e.g. an increase of electrode roughness or interface trap density. Ru has been further neglected as electrode material due to the tendency to form unfavorable RuO$_x$ at higher temperatures (see Section 4.1.2). Figure 7.9 b) shows leakage currents of capacitors with bottom and top electrodes stable at high annealing temperatures. Here, CaTiO$_3$ samples with TiN and Pt bottom electrode show comparable low leakage currents. The result can be explained by either a comparable barrier height between the bottom electrode and CaTiO$_3$ or a bulk dependent conduction process (e.g. Schottky emission). Obviously, the sample with carbon electrode exhibits a significantly higher leakage current, although the work function of carbon should be similar to the one of TiN. The differences may be explained by the surface oxidation of TiN as mentioned previously or carbon produces various interface defects with an effective reduction of the barrier height.

Although all samples possess the same TiN top electrode, all samples show different leakage currents for negative voltages. That can not be explained by surface roughness, hole conduction or barrier dependent conduction processes. The wet etching process of the TiN structuring may be responsible for the differences, as the capacitors with C and Pt bottom electrode lack of mechanical stability. This stability increases after annealing, visible in equal or lower leakage currents shown later in Figure 7.10 a). Nevertheless, as the leakage current at negative voltage follows the trend for the current at positive voltage, an influence of the electrodes on CaTiO$_3$ is expected to be independent of the barrier height.

Additionally, the use of Ru or TiN as both top and bottom electrode leads to asymmetric leakage currents for bottom and top electron injection. For Ru, the leakage current is higher, when negative voltage is applied to the top electrode. For TiN as top electrode, the leakage current is lower for top electrode electron injection. This asymmetry has been reported for various stacks in literature, like Pt/SrTiO$_3$/Pt [147], TaN/SrTiO$_3$/TaN stacks [185], RuO$_2$/TiO$_2$/RuO$_2$ [197], Ru/TiO$_2$/Ru [49] and TiN/

---

^{2}RuO$_2$ has a higher work function compared to Ru, which is beneficial for the reduction of leakage currents. For TiN electrode, the oxidation forms a lower k dielectric, which may be beneficial due to a thicker effective dielectric and higher band offset, but a reduction of capacitance density is most likely.

$Zr_{1-x}Al_xO_2$/TiN [113, 198]. Influence of the top electrode etching process can be excluded in this work due to the analysis of capacitors with different areas/parameters (Section 3.1), which should reveal any size dependent current behavior. Therefore, the asymmetry of leakage currents implies a difference in the top electrode/CaTiO₃ and CaTiO₃/bottom electrode interfaces. In Ref. [113], the asymmetry is the result of an oxidation of the TiN bottom electrode during deposition of ZrO_2. There, the top electrode does not show signs of oxidation, because the top electrode was made by depositing TiN on ZrO_2 without oxygen containing atmosphere. In this work, an interface formation can also appear for the sputter deposition of CaTiO₃ on top of Ru bottom electrode with an interfacial oxidation of Ru forming RuO_2. With the higher work function of RuO_2 and corresponding higher band offset, less electrons may be injected through the bottom electrode into CaTiO₃. While even the thinnest interface layer for both bottom electrode and top electrode can never be excluded, hints of an interfacial RuO_2 layer at the electrode/CaTiO₃ interface has been observed in TEM images, e.g. Figure 6.1).

The difference in leakage currents may further be assigned to different roughnesses of both bottom electrodes [199]. The influence of electrode roughness on leakage currents had been investigated in literature via simulations [69] and experimentally [200]. With increased roughness of the electrode-dielectric interface, the higher local field enhancement results in a large leakage current. In order to systematically investigate the effect of roughness, a bottom electrode material with a work function comparable to TiN and a significantly lower roughness has been searched for. Pyrolytically deposited carbon electrodes were chosen as they exhibit the lowest roughness of all electrodes (Section 4.2). As shown in Figure 7.9 b), the roughness may not necessarily play an important role in the observed leakage currents, because samples with carbon exhibit an even higher leakage current compared to TiN.

Kim et al. [201] stated for RuO_x/SrTiO₃/TiN capacitors, that the conduction in SrTiO₃ is driven by trap-assisted tunneling (TAT) along oxygen vacancies V_O. The asymmetric distribution of V_O near the top electrode and the bottom surfaces is than responsible for the asymmetric leakage currents. This group also reported that the leakage current through SrTiO₃ is controlled by TAT with the same barrier height of 1.5 eV between the Fermi level of TiN or RuO_x to the conduction band of SrTiO₃ regardless of the metal work function due to strong Fermi level pinning. This TAT conduction would explain the increase of leakage current in CaTiO₃ at negative voltages for TiN top electrode and electrodes with different work function materials. A significant reduction of the leakage current at positive and negative voltages is therefore possible by lowering the amount of initial oxygen vacancies [201, 202].

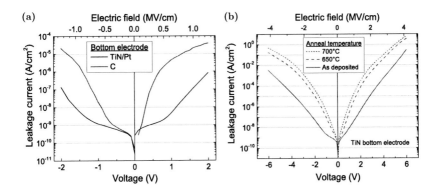

Figure 7.10: Leakage currents for annealed $CaTiO_3$ capacitors with optimized electrodes as a function of voltage. a) The capacitors with TiN/Pt and C bottom electrode (annealed at 650°C) are temperature stable (low increase in roughness, negligible interface/oxide formation) and exhibit a significant increase in permittivity. b) Capacitors with TiN bottom electrode exhibit a decrease of permittivity after annealing to a value of 20.

The temperature stable electrodes allowed a post-deposition anneal of the capacitors containing amorphous $CaTiO_3$. The leakage current of capacitors with those optimized electrodes and a $CaTiO_3$ thickness of 16 nm are shown in Figure 7.10 a). The currents differ for Pt and C electrode with an otherwise comparable capacitance of approx. 4×10^{-6} F/cm^2 for a crystalline $CaTiO_3$ layer. The differences in leakage currents between C and Pt electrodes at positive voltages can again be attributed to the different work functions. And the differences in leakage currents for negative voltages hint again to an intrinsic influence of the electrode to $CaTiO_3$. Samples with TiN bottom electrode exhibit an increased leakage current after anneal (Figure 7.10 b)). The measured capacitance density of these samples show low values comparable to the one in amorphous samples even after an annealing temperature of 700°C for 1 min in Nitrogen atmosphere. This and a positive quadratic voltage dependence of capacitance hint, that the $CaTiO_3$ layer is still amorphous or a large low-k interface layer is present. Therefore, investigations on crystalline $CaTiO_3$ samples where further done on samples with carbon and Pt bottom electrode.

To conclude the influence of different electrodes to the conduction behavior of $CaTiO_3$, the measurements show the lowest leakage current for capacitors containing amorphous

CaTiO$_3$ and a TiN or TiN/Pt bottom electrode with TiN top electrode. Capacitors with crystalline CaTiO$_3$ exhibit a leakage current of $1 \times 10^{-7} A/cm^2$ at 1 V for the samples with TiN/Pt bottom electrode together with a high permittivity. Samples with carbon electrode exhibit a high leakage current at large electric fields. Nevertheless, the excellent permittivity for those samples allows a reduction of leakage current within thicker CaTiO$_3$ layers, still matching the permittivity vs. leakage current requirements of the ITRS road map. It is also known, that introducing oxygen to the deposition or annealing process can reduce the defect density and therefore the leakage current to several orders of magnitude. The asymmetric distribution of oxygen vacancies near the top electrode and the bottom surfaces is supposed to be responsible for the asymmetric leakage currents for samples with the same bottom and top electrode.

7.3.2 Thickness dependence of leakage currents

The thickness dependent capacitance in CaTiO$_3$ MIM capacitors has been investigated with various electrodes and electrode arrangements (TiN, Ru, Pt, C) previously in Figure 6.6 and Figure 7.1. There, the capacitance is inversely proportional to the thickness for amorphous and crystalline CaTiO$_3$ capacitors. An influence of the interface to the capacitor's permittivity has not been observed in the thickness range of the investigated capacitors.

For leakage current measurements without an influence of the electrode/oxide interface, the leakage current should decrease with increasing the CaTiO$_3$ thickness. With normalizing the leakage current to thickness for different voltages, namely applying a similar electric field to the capacitor, the leakage current vs. electric field should be identical.

In Figure 7.11, the dependence of leakage current vs. the square root of the applied field for semicrystalline CaTiO$_3$ is presented. The $I(V)$ data has been previously published in Ref. [175] for the case of semicrystalline samples. There, the leakage current decreases significantly, when going from thin 10 nm to 30 nm oxide layers. The current reaches the resolution limit of the measurement unit at low voltages/applied fields. No thickness dependent behavior can be extracted below that limit. Above the resolution limit, the leakage current for the same electric field is identical for all different thicknesses and is not showing any field dependence. The mechanism valid for this conduction can be the Poole-Frenkel emission or tunneling into the conduction band. Other amorphous or crystalline CaTiO$_3$ thickness series' show comparable results.

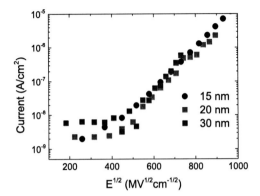

Figure 7.11: Leakage currents of various thicknesses of semi-crystalline CaTiO₃ capacitors with Pt bottom an Ru top electrode deposited at 550 °C as a square root function of the applied electrical field. The leakage currents decrease with increasing thickness. In this plot, the different leakage currents are identical for different thicknesses, proving the same field dependence. The linear increase for the three thicknesses in this plot can be attributed to the Poole-Frenkel conduction mechanism in semicrystalline samples.

7.3.3 Temperature dependence of leakage currents

As mentioned briefly in the previous text, temperature dependent current voltage (IVT) measurements allow the differentiation of single leakage current mechanisms. This is possible due to the different dependence of the single current transport models from temperature and voltage. The knowledge of these models allows an understanding and further optimization of capacitors containing CaTiO₃. The equations for various conduction models are presented in Section 2.4. The extraction of barrier heights for the different conduction mechanisms out of IVT measurements is explained in detail in Appendix B.

Because temperature dependent $I(V)$ measurements are done manually, the experimental implementation is different to previously shown semi-automatic measurements and needs 'special' precautions. The settings for the most reliable extraction are briefly as follows:

- A first manual $I(V)$ measurement is done on a single capacitor. The reliability of

(a) Pt bottom electrode

(b) TiN bottom electrode

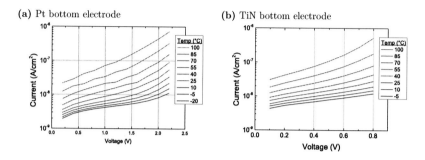

Figure 7.12: Temperature dependent $I(V)$ characteristics for amorphous CaTiO₃ on TiN and
Pt electrode with 16 nm oxide thickness ($T_{dep} = 350°$ C). The leakage current is
higher for the TiN electrode compared to the Pt electrode. a) The sample with
Pt electrode shows a more complex temperature dependence, indicating a change
of conduction mechanism at 2 V. The leakage behavior for high temperatures in
both samples is similar.

this single measurement has been tested on several capacitors on the same sample.

- The capacitor size is selected to ensure a measurement in the maximum possible
  voltage and temperature range with currents above the current resolution limit and
  below the compliance level of the measurement device.

- The sample is heated to the highest measurement temperature in a constant Nitrogen flow to desorb water from the surface. This prevents water condensation and
  the formation of ice crystals, when going to low temperatures below 0 °C.

- Next, measurements on a different capacitor are done at high temperatures, allowing the extraction of the maximum possible voltage (10-20% below the breakdown
  voltage).

- The measurement is done from high to low temperature.

- Due to the movement of the chuck, the capacitor's top contact has to be reset for
  every different temperature to ensure a good electrical contact.

- The delay time between ever voltage step has been set to 5 s in every measurement.

The temperature measurements are typically done between -40 °C and 120 °C. This range
shall allow a separation and verification of typical conduction mechanisms in the samples.
In Figure 7.12, IVT measurements for both Pt and TiN bottom electrode point out the

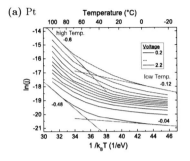 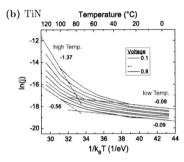

Figure 7.13: Arrhenius plots from Figure 7.12 for amorphous $CaTiO_3$ on Pt and TiN bottom electrode. The curvature is typical for IVT measurements with $CaTiO_3$ capacitors. For low temperatures, the gradient is strongly reduced (no temperature dependence) equally for small to high voltages. The gradient for high temperatures is increased significantly and show a strong voltage dependence. The highest gradient is extracted for high voltages and high temperatures.

typical behavior of $CaTiO_3$ capacitors. The samples have been deposited simultaneously at $T_{dep} = 350°C$ with a $CaTiO_3$ thickness of 16 nm. Both bottom electrodes have been selected due to the observation of at least two conduction mechanism changes in current for the applied voltage ranges. The curvatures are not convex, as it should be for single mechanisms e.g. thermionic emission and Poole-Frenkel conduction (they have a $\propto \exp(-\sqrt{V})$ dependence, which follow a convex curvature in the semi-log plot). Tunneling follows a $\propto \exp(-1/V)$ curvature. The concave behavior indicates a superposition of at least two mechanisms.

To distinguish between the different mechanisms involved, the leakage currents for the TiN and Pt electrodes used are plotted in dependence of temperature. The Arrhenius plots in Figure 7.13 show an at least quadratic shape for different temperatures for both electrodes. No single conduction mechanism can be derived out of the graph, because no constant slope is extractable in the measured temperature range. This curvature implies that at least two conduction mechanisms are contributing to the overall leakage current. Therefore, an estimation of involved conduction mechanisms is done using the boundary data points for the extraction of gradients (low temperature and high temperature region).

In various literature, Arrhenius plots with non-linear curvature had been observed previously, e.g. for Si_3N_4 MIM capacitors [203, 204, 205] or Ta_2O_5 [206]. Although the applied temperature range more than 100 K for both TiN and Pt bottom electrode, no sign of any linear dependence give a hint towards the contribution of more than two mechanisms. For low temperatures (largest $1/k_BT$), the gradient strongly diminishes and is constant for different voltages. The extracted gradient for TiN and Pt bottom electrodes lies at approx. 0.1 eV. This hints towards a temperature independent leakage current mechanism, namely tunneling. Different to low temperatures, the gradient in the Arrhenius plots is much larger at high temperature and increases with applied voltage. This implies a temperature activated leakage current mechanism, for example Poole-Frenkel or thermionic emission.

Additionally, from the Arrhenius plots Figure in 7.13, it is possible to extract the voltage dependent activation energies $E_A(V)$ from the slope of the leakage currents at the boundaries at high and low temperatures. This extraction is demonstrated for a single conduction mechanism in the Appendix B. As it is shown here or in Figure 2, the activation energies vary with voltage at a specific temperature. An extrapolation of the $E_A(V)$ function to zero voltage (Equation (7.9)) results in the barrier height ϕ_B of the temperature dependent conduction mechanism according to Equation (7.9).

$$\ln I \quad \propto \quad -E_A(V) \times \frac{1}{k_BT}$$
$$E_A(V=0) \quad = \quad \phi_B \qquad\qquad (7.9)$$

For the different temperatures, the activation energies $E_A(V)$ for both Pt and TiN electrodes are plotted versus voltage in Figure 7.14. For high temperatures, the change in E_A with voltage is much more pronounced compared to E_A of lower temperatures. The higher temperature might activate temperature dependent emission of electrons from traps in $CaTiO_3$. This process is suppressed at temperatures below 20 °C in Figure 7.13. The barrier heights are extracted with extrapolation to zero voltage (Equation (7.9)). An interpretation of the different barrier heights is rather difficult. Deeper traps between 0.8-1.2 eV have been related to oxygen deficiency (e.g. Ref. [75]), while barrier heights of 0.4-0.6 eV have been related to structural defects in the amorphous oxide. Both explanations are reasonable for the extracted barrier heights of the TiN or Pt containing capacitor stack in Figure 7.14.

An even more complex behavior for temperature dependent leakage current measurements is shown in Figure 7.15 for another capacitor stack containing Pt as electrode. The amorphous $CaTiO_3$ layer with a thickness of 11 nm has been deposited at 450 °C. To accomplish a reduced defect concentration in the oxide, the higher deposition temperature compared to the previous examples should result in a reduced defect concentration

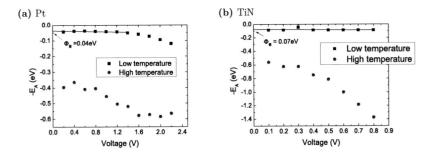

Figure 7.14: Activation energy as a function of voltage extracted from the data of Figure 7.13 for amorphous CaTiO₃ on Pt and TiN bottom electrode. The activation energies for low temperatures are close to zero and show a minor dependence on voltage, while high temperature activation energies exhibit a strong voltage dependence.

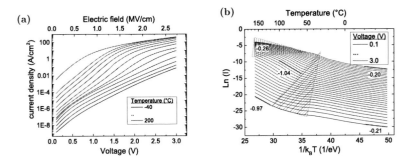

Figure 7.15: Temperature dependent $I(V)$ measurement (IVT) of 11 nm amorphous CaTiO₃ in a Pt/CaTiO₃/TiN capacitor (T_{dep} = 450 °C). a) The IVT measurement reveals a rather complex curvature with at least 3 conduction mechanisms involved. b) The Arrhenius plot deduced from data of a) shows two regions with voltage independent activation energy (red lines). These are divided by transition regions (marked net). At high voltages and temperatures, the current saturates (gradient of -0.26 eV).

and therefore less defect driven conduction. Nevertheless, at least three different conduction mechanisms can be attributed to the curvature following a convex-concave-convex transition ($I(V)$ at 60 °C) in Figure 7.15. For high temperatures and high voltages, an ohmic behavior independent of voltage and temperature is expected according to Ref. [70]. At low temperatures, the $I(V)$ curvature implies a square root dependence from voltage.

The Arrhenius plot in Figure 7.15b) underlines the ohmic behavior with no voltage dependence at high temperature and high voltage (gradient of -0.26 eV). At low temperatures, a barrier height ϕ_B of 0.2 eV can be extracted from the constant slopes in the Arrhenius plot. This low barrier height for low temperatures has been attributed in literature to a tunnel-assisted transfer between neighboring trap centers [207]. The mid region part of the Arrhenius plot reveals an activation energy or barrier height of about 1 eV, typically assigned to ionic polarization [207] and/or oxygen vacancy traps in the oxide. The linear slopes in the Arrhenius plot for low temperatures suggest a Poole-Frenkel conduction. Because the Poole-Frenkel conduction is temperature dependent, a Poole-Frenkel emission process especially at low temeperatures seems to be unlikely.

A different part of the same sample has been crystallized at 600°C and measured temperature dependent (Figure 7.16). The CaTiO₃ is known to be completely crystallized from capacitance measurements (e.g. in Figure 7.6) as well as TEM pictures (e.g. Figure 6.3). The complex leakage current behavior for the amorphous CaTiO₃ capacitor in Figure 7.15 has been significantly simplified. The Arrhenius plot shows a linear slope proportional to the inverse temperature (Figure 7.16a)). In contrast to the amorphous sample, this allows the extraction of the voltage dependent activation energy. The E_A dependence on the voltage is shown in Figure 7.16b). For low voltages, the extrapolation to zero voltage results in $E_A(0) = \Phi_B$ =(0.75-0.85) eV. The extraction of barrier height is equal for capacitors measured in different temperature regions as well as positive and negative voltages. Due to the asymmetric electrode arrangement in the MIM capacitor stack with Pt as bottom electrode and TiN as top electrode, the constant barrier height can be assigned as an intrinsic property of crystalline CaTiO₃ , namely the trap depth of 0.8 eV for Poole-Frenkel emission of electrons. This trap energy has been assigned to oxygen deficiencies in HfO₂ in literature [75].

Summarizing the previous results of temperature dependent leakage currents on CaTiO₃ capacitors, no specific conduction mechanism can be assigned with apodeictic certainty. In ultrathin amorphous CaTiO₃ capacitors, several conduction mechanisms contribute to the overall leakage currents. It is obvious, that amorphous layers of sputtered CaTiO₃ contain various defects and therefore defect driven processes are responsible. The assump-

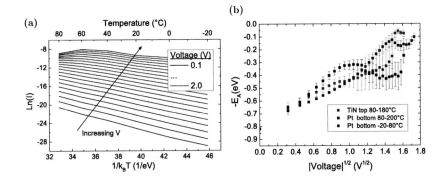

Figure 7.16: Temperature dependent $I(V)$ measurement of 11 nm crystalline CaTiO$_3$ in a Pt/CaTiO$_3$/TiN capacitor ($T_{an} = 600\,°C$). a) The Arrhenius plot shows a linear dependence with inverse temperature. The measured temperature range for the capacitor is -20 - 80 °C and injection of electrons from the Pt electrode. b) Extracted slope from a) as activation energy with results from measurements of different capacitors and temperature regions and voltages. Extrapolation to zero voltage reveals a barrier height of $\Phi_B \approx 0.8\,eV$ for all capacitors within the same stack.

tion of a single trap level in the conduction process may be replaced by a distribution of trap levels. Or an interaction of traps may occur [208]. Furthermore, the non-linearity can be the result of a transition at high fields to trap assisted tunneling, at low fields to hopping conduction and at high fields and low temperature to Fowler-Nordheim conduction [209]. All these assumptions can be valid for the conduction process in amorphous CaTiO$_3$. For crystalline CaTiO$_3$ capacitors, the conduction process shows a linear dependence with (inverse) temperature. Only one conduction mechanism is the dominant process responsible for this leakage current. The extracted barrier height is $\Phi_B \approx 0.8\,eV$. This is a typical value, when oxygen vacancies are present in the dielectric. Therefore, further steps to improve the leakage current of crystalline CaTiO$_3$ should include processes to reduce the defect density, i.e. by the introduction of oxygen during deposition to reduce the oxygen deficiency.

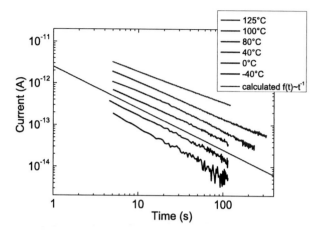

Figure 7.17: Temperature dependence of relaxation currents in a capacitor with 30 nm crystalline CaTiO$_3$. The layer has been deposited at 600°C on top of a carbon electrode. The slopes vary minimally around the ideal Curie-von Schweidler power law behavior (t^{-1}) (blue line).

7.4 Relaxation

Despite the measurement of leakage currents in CaTiO$_3$ capacitors, relaxation currents are always present in a dielectric (Section 2.5). All leakage currents measured previously are done with a delay time of 5 s between each measurement step to reduce the influence of these relaxation currents. In Figure 7.17, the current decay versus time is shown for different temperatures on a crystalline capacitor with a 30 nm thick CaTiO$_3$ layer. All currents follow the Curie-von Schweidler law with slopes in the order of minus one. The slight temperature dependence of the magnitude of the relaxation currents indicates no typical thermally activated processes [82]. The results are comparable to results of Ba$_x$Sr$_{1-x}$TiO$_3$ [210].

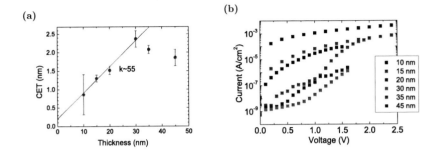

Figure 7.18: Electrical results for thickness series. On Pt electrode, the k-value of 55 is extracted up to a thickness of 30 nm (Figure 7.19a-d)). The further increase to 104 indicates a partial crystallization of the capacitor on thinner layers. The corresponding leakage currents reveal a reduction of currents for layers up to 30 nm and an unexpected significant increase for thicker layers.

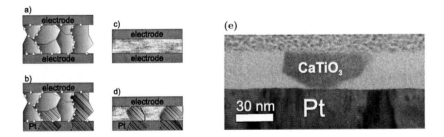

Figure 7.19: Various crystal states of CaTiO$_3$ and the preferential growth of isolated crystalline CaTiO$_3$. a-d) show the various crystal states in a CaTiO$_3$ capacitor stack. a,b) show the fully crystalline state for thick CaTiO$_3$ layers without and with an order correlation to the underlying electrode. c,d) depicts the crystallization state for thin CaTiO$_3$ layers. c) The crystallization is prevented in the thin oxide due to the increase of crystallization temperature after Equation (6.1) (Chapter 6). d) With preferential growth on specifically ordered Pt electrodes, isolated crystals grow in an otherwise amorphous matrix. e) The TEM micrograph shows an isolated CaTiO$_3$ crystal in an amorphous matrix.

7.5 Optimized capacitance and leakage using semicrystalline samples

It has been shown in the previous chapter, that with the increase of deposition temperature to 550 °C, CaTiO$_3$ starts to crystallize. The permittivity of different capacitors on a Pt bottom electrode increased to a value of k above 50. In order to study this in detail, a thickness series for Pt electrodes at the deposition temperature of 550 °C was investigated, while all other deposition conditions are kept constant (Figure 6.6). In Figure 7.18a), a total thickness versus CET plot is used to depict the development of permittivity with the thickness. The k-value can be extracted from the slope of a linear fit for several data points. Layers between 10-30 nm thickness show a constant k-value of 55, while for thicker samples, the k-value increases up to 105. From TEM images (e.g. Figure 6.7), the lower permittivities can be associated to a partial crystallization of the thinner samples. According to Figure 7.18a), the constant slope indicates a constant k-value and therefore a constant degree of crystallinity for the thin samples. The intercept of the linear fit at a CET of 0.25±0.18 nm corresponds to a negligible interfacial layer between the CaTiO$_3$ and the bottom electrode. This result completes the results from amorphous layers in Figure 7.1 for semicrystalline samples.

In Figure 7.18b), the corresponding leakage currents for CaTiO$_3$ layers deposited at 550 °C with different layer thicknesses are plotted. Low leakage currents comparable to amorphous layers (e.g. Figure 7.9) are shown for the samples between 15 nm (2.0 × 10^7 A/cm^2 @1V) and 30 nm (3.57 × 10^{8}A/cm^2 @1V) layer thickness. According to a Poole-Frenkel plot in Figure 7.11, the leakage current here can be assigned to Poole-Frenkel (PF) conduction. Unexpectedly high leakage currents are measured for samples with a layer thickness above 30 nm even for low voltages. Trap assisted tunneling along many defect states at grain boundaries is suggested to be the major leakage mechanism. The strong increase in leakage current for the 10 nm sample can be attributed to the rough surface of the bottom electrode of the thinnest film.

The increase of crystallinity correlates well with the increase in leakage current for the thicker samples. For the 10-30 nm thick layers, the k-value of 55 is a result of a small number of crystallites with a high k-value embedded in an amorphous layer with a smaller k of 26. The constant k-value of 55 implies a constant degree of crystallinity with changing the oxide thickness up to 30 nm. Distinctly, the leakage currents are comparable to amorphous samples. The degree of crystallization observed in Figure 7.18 is sketched in Figure 7.19a-d), where the layer develops from isolated CaTiO$_3$ crystals (Figure 7.19 d)) to a fully crystallized CaTiO$_3$ stack (Figure 7.19 b)). This result strongly suggests, that the grain boundaries as main leakage paths [86, 87] of the isolated crystallites are passivated by the amorphous matrix, as it has been shown for other materials by Heitmann et al. [211]. In Figure 7.19e), a single $CaTiO_3$ crystallite in an otherwise amorphous matrix is shown. The reasons for the preferential growth of the crystallites have been discussed extensively in Chapter 6.

The higher conduction through layers with higher degree of crystallinity has been attributed to the conduction along grain boundaries (sketched electrons in Figure 7.19 a) and b)). Planar faults in the $CaTiO_3$ crystals due to stoichiometric variations have not been observed in TEM micrographs (e.g. Figure 6.2) as shown by Gu and Ceh [212], resulting in an increase of leakage current through the crystallites. This implies that the formed $CaTiO_3$ crystals reach the perovskite stoichiometry locally. The current conduction through a single crystallite is limited to structural defects, vacancies, or grain boundaries. With these restrictions, leakage current discrepancies between thin and thick oxide layers can be primarily attributed to a large increase in the number of grain boundaries for layer thicknesses above 30 nm without passivation by an amorphous matrix. This leads to a much higher number of leakages paths along grain boundaries [86, 87]. As a result, with oxide thicknesses greater than 30 nm, leakage current increases significantly.

The results presented in this section (published in Ref. [175]) allow to search and investigate samples with partially crystalline $CaTiO_3$ in an amorphous matrix. The preparation process is depicted schematically in Figure 7.20. Annealing the samples after deposition may simplify the search for capacitors with partially crystallized $CaTiO_3$ layers.

7.6 Conclusion

In this chapter, the electrical properties of amorphous and crystalline $CaTiO_3$ capacitors have been investigated in detail. The experiments reveal the properties of $CaTiO_3$ dependent on preparation variations and different electrodes. The capacitance voltage measurements showed k-values of about 100 for crystalline layers even for the thinnest prepared layers with 10 nm thickness. Between both the amorphous and the fully crystalline states exist a variety of crystal states with partially crystallized $CaTiO_3$. The reason for the appearance of the intermediate state has been investigated physically in the previous chapter and their influence to leakage currents is discussed here extensively. A non-linear voltage dependence of capacitance, which is immanent for perovskites with high permittivity. These non-linearities of $CaTiO_3$ have been depicted in detail, especially the quadratic VCC. The dependence of the quadratic VCC has been evaluated for

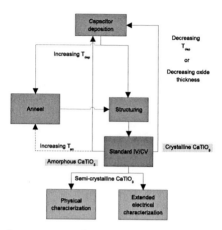

Figure 7.20: Scheme for preparation of semi-crystalline $CaTiO_3$ samples. The electric measurement is necessary to evaluate the permittivity of the $CaTiO_3$ layer. XRD measurements are incapable of extracting crystal $CaTiO_3$ peaks due to low crystal density and texturing of the oxide. Other electrodes have been tested, but only the use of Pt electrodes $CaTiO_3$ shows the semi-crystal state.

different temperatures, electrodes, frequencies and thicknesses. The quadratic VCC of $CaTiO_3$ is insignificantly small compared to values for $SrTiO_3$ and $BaTiO_3$ as competing high-k materials. Even at high electric fields of $1\,MV/cm$, the drop of $CaTiO_3$ permittivity is very low compared to values from literature [56]. Nevertheless, the VCC of $CaTiO_3$ is significantly larger than the limits ($< 100 ppm/V^2$) necessary for the integration in RF devices. Various methods are available to further reduce the VCC to reach the required limits, like oxygen vacancy reduction or the introduction of bilayers. The leakage currents for the prepared capacitors, summarized in Figure 7.21, reveal challenging properties to fulfill the requirements for dielectrics of future DRAM capacitors [184]. The thinnest layers prepared with a CET of $0.49\,nm$ show leakages currents of $6.11 \times 10^{-7}\,A/cm^2$ at $1\,V$ (Figure 7.21). $CaTiO_3$ samples with a CET of $0.75\,nm$ and a leakage current below $1 \times 10^{-8}\,A/cm^2$ at $1\,V$ were shown. The advantage of the high band gap of $4.0\,eV$ from low energy EELS measurements in Chapter 5.3 allows the use of capacitors with TiN and carbon electrodes. With carbon as bottom electrode, the capacitors show leakage currents of $6.8 \times 10^{-9}\,A/cm^2$ with a CET of $1.13\,nm$, mainly due to the superior low roughness and stability of the carbon layer.

In these experiments, no process to reduce the oxygen deficiency of the $CaTiO_3$ layers

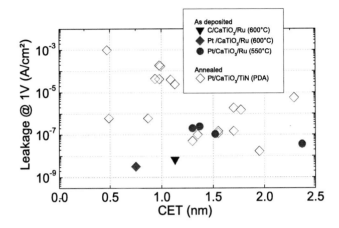

Figure 7.21: Overview of leakage currents versus CET of crystalline CaTiO₃ capacitors. Improvements to electrode roughness and deposition process allow leakage currents below 1×10^{-7}A/cm² for crystalline CaTiO₃ capacitors.

was applied. For crystalline samples, a defect depth of 0.8 eV attributed to oxygen vacancies could be extracted. An addition of oxygen during the deposition of CaTiO₃ could further reduce the defect density and with this the contribution of Poole-Frenkel emission or other trap related conduction processes to the overall leakage current.

8 Summary and Conclusion

This dissertation is an unprecedented investigation on the deposition, electrical characterization and physical properties for nanometer thickness $CaTiO_3$ high-k dielectrics in charge-storage capacitors. Hereby the efforts rely on the development of capacitors that fulfill the demands of the 2012 ITRS road map for DRAM capacitors for 2016 and beyond, enabling further downscaling of electronic devices with higher capacitance at low power consumption. The requirements for DRAM are a CET of 0.50 nm ($t_{phys} = 7.7nm$) in 2016, a k-value of 70 till 2024 and a leakage current minimum of 1.079×10^{-7} A/cm^2 at 0.6 V. These are stringent requirements that make the study of new materials in this field challenging. Profound scientific studies are required to assess the potential of $CaTiO_3$ for future DRAM applications. This thesis encompasses the full process and development of high temperature stable electrodes, the capacitor fabrication and characterization.

The $CaTiO_3$ capacitors were deposited by sputtering in a customized UHV sputter system, allowing high purity layers without the inclusion of carbon and other impurities into the oxide. Using the connected metal chamber, the deposition of the whole capacitor stack without breaking the vacuum ensured the exclusion of adsorbents at the interfaces. Various electrodes have been tested for suitability and stability during high temperature depositions to achieve crystalline $CaTiO_3$ layers. With this setup, ultrathin $CaTiO_3$ layers in a working capacitor stack were produced and studied both physically and electrically to address the requirements in future electronic devices. A novel approach of reducing the leakage currents in $CaTiO_3$ capacitors has been investigated, where nanocrystals of $CaTiO_3$ were surrounded by an amorphous $CaTiO_3$ matrix. The leakage currents match the values for amorphous layers by the passivation of grain boundaries with the advantage of an increased k-value compared to amorphous samples.

Deposition of temperature stable bottom electrodes

The bottom electrodes were designed to withstand the stringent requirements for the complete stack and achieve the crystallization temperature of $CaTiO_3$. These requirements are the integrity and exceptionally low surface roughness, the absence of interfacial layers detrimental to the overall capacity and suppression of intermixing upon $CaTiO_3$ deposition. TiN as standard electrode material used in industry tends to oxidize during deposition at temperatures above 400 °C hindering the use as bottom electrode. Pt as electrode on the other hand requires a diffusion barrier to prevent the formation of Pt silicides. While several approaches in literature use thick SiO_2 or Al_2O_3 oxide layers as barrier, they are not conductive and ultimately limit the scalability in a vertical capacitor. To this end, TiN and RuO_2 have been investigated as conductive barriers, TiN was found to be the only electrode material that fits the requirements after high temperature processes. Different to Pt electrode preparations in literature, in this work the Pt layers have been optimized to a thickness of only 20 nm as Pt exhibits high material costs.

As alternative to Pt, a CVD grown pyrolytic carbon layer was found to be a suitable temperature stable bottom electrode for achieving crystalline $CaTiO_3$. It exhibits high stability versus temperature and chemical reactions and can easily be structured by oxygen etching steps. In literature, carbon had recently been introduced as conformal electrode in DRAM trench capacitors showing its suitability to fit in industrial processes. The convenient results in leakage currents of the $CaTiO_3$ capacitor stacks indicate a sufficient band offset of TiN and carbon electrodes to $CaTiO_3$ at lower voltages. This may allow the complete replacement of noble metal electrodes like Ru and Pt. Differences in leakage currents between TiN and carbon have been attributed to the roughness of these electrodes after the $CaTiO_3$ deposition at elevated temperatures. Further improvements to the temperature stability of electrodes may ultimately result in superior leakage currents of thinnest $CaTiO_3$ layers.

Physical characterization of $CaTiO_3$ layers

To evaluate the electrical results of $CaTiO_3$, intensive studies of physical properties of $CaTiO_3$ layers are necessary to classify the sputter deposition process. The sputtered layers of $CaTiO_3$ have been investigated regarding the optical properties using UV-vis spectral ellipsometry. The values for refractive index and extinction coefficient are lower compared to literature values of $CaTiO_3$ bulk crystals due to lower density of sputtered layers. The extracted band gap is 4.0-4.2 eV measured with ellipsometry and low-energy electron energy-loss spectroscopy. The high purity $CaTiO_3$ layers are accountable for the larger band gap compared to literature values of $CaTiO_3$. This is a distinct advantage to achieve low leakage currents compared to $SrTiO_3$ with a band gap of only 3.2-3.5 eV. XPS measurements revealed the existence of adsorbents at the $CaTiO_3$ surface corresponding to CaO, CaOH, $CaCO_3$ formation. This result supports the hygroscopic nature of $CaTiO_3$ and a reaction with ambient gases. The *in-situ* deposition of the full capacitor stack is therefore a necessary step to achieve the maximum permittivity of pure $CaTiO_3$ films without the formation of any low-k interfaces. Stoichiometry measurements with EELS and XPS revealed a Ca/Ti ratio of 1.0 for the sputtered layers. An oxygen deficiency in the layers has been measured, however no significant influence to leakage currents due to e.g. trap-band formation is visible.

The crystallization temperature is, besides the oxide thickness and the deposition/ annealing temperature, influenced by the choice of substrate. With HTXRD measurements, the crystallization temperature of amorphous $CaTiO_3$ (30 nm) on Si substrates has been determined to be 640 °C. A starting crystallization of $CaTiO_3$ has been observed for samples with Pt as bottom electrode deposited at 550°C and with rapid thermal annealing at 600°C .

Selected samples have been investigated with TEM to gain knowledge of the crystallinity for maximum permittivity of $CaTiO_3$ in thin films. The reason for intermediate values of permittivity has been correlated to partially crystallized $CaTiO_3$ layers. The formation of crystallites is substantially influenced by the Pt bottom electrode. There is a preferential growth of $CaTiO_3$ crystals on top of a Pt (111) surface resulting in isolated crystals in an otherwise amorphous matrix. The influence of the isolated crystals on electrical properties has been addressed in Ref. [175]. Going to higher deposition/annealing temperatures and layer thicknesses, the layers become fully crystallized and the preferentially oriented crystallites are overshadowed by the formation of unordered crystallites.

The use of electrodes which have a major deviation in lattice constants from $CaTiO_3$ would result in strain induced to the dielectric lattice. Due to the good lattice match of Pt towards $CaTiO_3$, no lattice strain have been observed regarding this electrode. While the influence of lattice strain towards leakage currents is not known, it may inhibit the maximum permittivity of the overall $CaTiO_3$ layer.

Electrical performance of $CaTiO_3$ capacitors

With detailed physical analysis of the prepared $CaTiO_3$ capacitor stacks, the influence to electrical properties of amorphous and crystalline $CaTiO_3$ capacitors has been intensively investigated. With the used UHV sputter system, the contaminations of $CaTiO_3$ have been reduced to a minimum. The oxygen deficiency of the sputtered $CaTiO_3$ layers do not show to have an influence on the leakage currents (e.g. by trap band formation). Nevertheless, the reduction of oxygen deficiencies in the $CaTiO_3$ layers may further reduce the leakage currents. Impedance spectroscopy measurements in the range of 40 Hz to 100 MHz revealed no significant deviation of amorphous, partially and fully crystalline CaTiO$_3$ capacitors from an ideal capacitor. The relaxation currents in CaTiO$_3$ capacitors follow the Curie-van Schweidler law similar to SrTiO$_3$ and other high-k materials investigated in literature. Together with the impedance measurements, the results show the potential of ultrathin CaTiO$_3$ layers in RF applications.

The capacitance voltage measurements showed k-values of about 100 for fully crystalline layers even for the thinnest prepared layers with 11 nm thickness. The full evolution from amorphous to crystalline CaTiO$_3$ samples are summarized in Figure 8.1. Between both amorphous and fully crystalline CaTiO$_3$ layers, a variety of intermediate permittivities exists. As known from TEM measurements, these intermediate permittivities are the result of layers containing partially crystallized CaTiO$_3$.

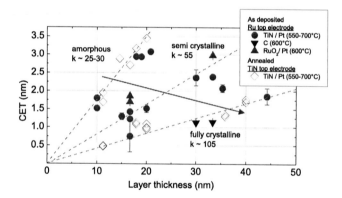

Figure 8.1: CET as a function of layer thickness for different degrees of crystallinity and k-value of CaTiO$_3$ in various capacitor stacks. As a general trend, with the increase of the deposition/annealing temperature, the permittivity increases from 25-30 for amorphous to ≈55 for semi-crystalline and ≈105 for fully crystalline CaTiO$_3$ capacitors. Details can be found in Chapter 7.

Capacitance measurements reveal a non-linear voltage dependence of capacitance, immanent for perovskites with high permittivities. The values and dependence of these non-linearities of CaTiO$_3$ have been described in detail, namely the quadratic voltage capacitance coefficient (VCC). The dependence of the quadratic VCC has been shown for different temperatures, electrodes and thicknesses. The quadratic VCC is significantly lower compared to thin film capacitors containing SrTiO$_3$ and Ba$_x$Sr$_{1-x}$TiO$_3$ as competing high-k materials. At high electric fields of $1\,MV/cm$, the drop of permittivity of $CaTiO_3$ capacitors is so low that these capacitors exhibit an even higher permittivity compared to the high-k materials mentioned previously [56]. Various methods are available to further reduce the VCC of crystalline $CaTiO_3$ to enable the integration in RF applications. The measurements of leakage currents reveal the intriguing properties

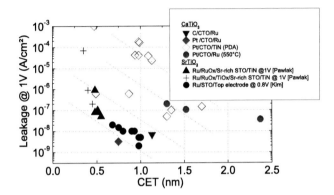

Figure 8.2: Performance of $CaTiO_3$ capacitors compared to literature values of $SrTiO_3$ capacitors (Pawlak et al. [202]; Kim et al. [1]).

of $CaTiO_3$ compared to $SrTiO_3$. The leakage currents of $CaTiO_3$ at a certain CET are comparable to highly optimized $SrTiO_3$ capacitors (depicted in Figure 8.2). The thinnest layers prepared with a CET of $0.49\,nm$ show leakages currents of $6.11 \times 10^{-7}\,A/cm^2$ at $1\,V$. $CaTiO_3$ samples with a CET of $0.75\,nm$ and a leakage current below $1 \times 10^{-8}\,A/cm^2$ at $1\,V$ even surpass the results of $SrTiO_3$ capacitors. Beside the astonishing results of leakage currents of capacitors containing high work function Pt electrodes, the advantage of the high band gap of $4.0\,eV$ allows capacitors with $CaTiO_3$ on carbon electrodes. These capacitors show leakage currents of $6.8 \times 10^{-9}\,A/cm^2$ with a CET of $1.13\,nm$ corresponding to a permittivity of 105. Further scaling the oxide thickness will allow the values to lie below the thickness limits of $8\,nm$ demanded by industry. To conclude the performance of $CaTiO_3$ MIM capacitor stacks, $CaTiO_3$ exhibits k-values of approximately 100 with exceptionally low voltage dependence and low leakage currents on Pt as well as carbon electrodes, already fulfilling the stringent requirements for future DRAM applications with the parameters investigated in this work. Further improvements to electrode roughness and vacancy reduction can allow scaling to thinner $CaTiO_3$ layers with thicknesses below $8\,nm$ with targeted leakage currents to allow the integration in recent structures.

Extensive electrical characterization of reliability may prepare the ground for the integration in DRAM capacitors. As future work, the leakage currents of $CaTiO_3$ can be further reduced by the introduction of dopants like Al and Y for defect management. As a valuable academic research topic, the preparation of nanocrystals in an amorphous matrix can be extended and tested on other perovskite materials.

A Extraction of optical constants

The relation between refractive index n (and extinction coefficient κ) and relative ε_r is

$$n = \sqrt{\varepsilon_r \mu_r} \approx \sqrt{\varepsilon_r} \qquad (A.1)$$

μ_r is the relative permeability. For most materials, μ_r is close to 1 at frequencies in range of the visible light and therefore n is approximately $\sqrt{\varepsilon_r}$. ε_r corresponds to the permittivity in the same frequency range and is typically close to ε_∞ (the electronic part of the dielectric function). At optical frequencies, absorption occurs in most dielectrics and ε_r becomes complex. The imaginary part ε_2 of the complex dielectric function $\tilde{\varepsilon}$

$$\tilde{\varepsilon} = \varepsilon_1 + i\varepsilon_2 \qquad (A.2)$$

The correlation between ε_1 and ε_2 and refractive index n and extinction coefficient κ is given as

$$\varepsilon_1 = n^2 - \kappa^2 \qquad (A.3)$$

$$\varepsilon_2 = 2n\kappa \qquad (A.4)$$

The interpretation of the measured ellipsometric parameters Ψ and Δ require models to extract the optical constants. For materials, which are transparent ($\varepsilon_2 = 0$ or $\kappa = 0$) in the measured visible spectrum, the Cauchy equation allows the determination of the refractive index. It consists of an empirical relationship between the refractive index $n(\lambda)$ and wavelength of light λ (Equation (A.5))

$$n(\lambda) = A + \frac{B}{\lambda^2} + \cdots \qquad (A.5)$$

,where A, B are Cauchy parameters. When reaching regions of anomalous dispersion, in particular absorption edges of the material, this model is inaccurate.

A simple model to describe the adsorption ε_2 at the band gap (the adsorption edge) is the Tauc formular in Equation (A.6). The imaginary part of the dielectric function near the band edge is described according to Ref. [213]

$$\varepsilon_2(E) = A_T \frac{(E - E_g)^2}{E^2} \Theta(E - E_g) \quad . \qquad (A.6)$$

E is the photon energy, E_g is the band gap of the material, A_T is a constant, and Θ is the Heaviside Theta function with $\Theta(x < 0) = 0$ and $\Theta(x > 0) = 1$.

A state-of-the-art model for the description of the complex refractive index around absorption edges is the Tauc-Lorentz model (Equation (A.7)). It uses the expression of Tauc and combines it with the complex dielectric function for a Lorentz oscillator [214].

$$\varepsilon_2 = \frac{AE_0\Gamma\ (E - E_g)^2}{(E^2 - E_0^2)^2 + \Gamma^2 E^2}\ \frac{1}{E}\ \Theta(E - E_g) \tag{A.7}$$

The formula for the imaginary part of the dielectric function contains the band gap E_g, the peak transition energy E_0, the broadening parameter Γ and A, which represents the optical transition matrix elements. The real part ε_1 is than obtained using the Kramers-Kronig relation

$$\varepsilon_1 = \varepsilon_\infty + \frac{2}{\pi} \int_0^{+\infty} \frac{\xi\varepsilon_2(\xi)}{\xi^2 - E^2} d\xi. \tag{A.8}$$

Using the Tauc-Lorentz oscillator model works well for $CaTiO_3$ as well as other dielectrics in the UV-visual range [215].

B Investigation of temperature leakage currents and extraction of barrier heights

The measurement of temperature dependent leakage currents (IVT) shall provide information of the leakage mechanism involved in the conduction process. Several models for leakage currents are available, e.g. Schottky/thermionic emission, Poole-Frenkel conduction, Trap-assisted tunneling and Direct or Fowler-Nordheim tunneling. To gain access to these specific models, the IVT measurement is a suitable procedure to differentiate between all these models, because every model has a different dependence on voltage and temperature. The most general equation for the given models is

$$j(V,T) = A(V,T) * e^{E(V,T)} \tag{B.1}$$

with $A(V,T)$ as the pre-exponential factor with the unit of current and including the density of available charge carriers in the insulating layer and $E(V,T)$ as the function for the carrier barrier behavior typically proportional to $1/k_B T$ (k_B - Boltzmann constant, T - temperature).

With the Equation (B.1), several graphical representations allow to specify the different conduction models. The main goal for these charts is the possibility to linearize the measurement data. The graphical analysis is easier compared to the simulation of countless fit functions. The graphic charts are named after the conduction process one wish to extract (e.g. Poole-Frenkel or Schottky plot for voltage dependence or Arrhenius plot for temperature dependence).

To start with, a series of leakage currents over voltage and temperature has to be measured (or simulated). A simulated temperature dependent leakage current series is shown in Figure 1. The curvature is comparable to some measured data for leakage currents. The point of intersection describes the voltage, where the the barrier height is finally zero (never reached in a real measurement, will be supplanted by tunneling current). With $E(V,T) = 0$, the leakage current finally reaches A. Even with the simple equation, the

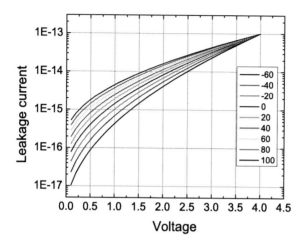

Figure 1: Simulated leakage currents after Equation (B.1) with constant A. Parameters: $A = 1 \times 10^{-13}$, $E(V, T) = -(\Phi_B - 0.1 \times \sqrt{V})/k_B T$ with $\Phi_B = 0.2$ and $T \in [-60°C, 100°C]$. The pre-exponential factor is here constant for simplicity reasons, the exponential function $E(V, T)$ is comparable to Schottky or Poole-Frenkel emission behavior.

real curvature is unknown and difficult to fit.

To simplify the view on the data series, the leakage current can be shown in an Arrhenius plot in Figure 2. The Arrhenius graph plots the leakage current over the inverse temperature $1/k_B T$. The linearity of all lines in this graph means that $E(V, T)$ is proportional to $1/k_B T$ and the pre-exponential factor $A(V)$ is independent of temperature. Nothing can be told about the voltage dependence of leakage current from the Arrhenius plot. Any change in $A(V)$ results in a constant vertical shift in the Arrhenius plot.

To further investigate the voltage dependence of $E(V, T)$ and extract the physically relevant barrier height, the gradient of every line is plotted over the voltage as seen in Figure 3a. The graph shows the voltage behavior of $E(V)$ for all voltages. $E(V)$ is now independent of temperature because of the calculation of the gradient from the linear fits for all temperatures in the Arrhenius plot in Figure 2. To extract the barrier height Φ_B out of the $E(V)$ plot, the graph has to be extrapolated to $E(0) = \Phi_B$. For the given example, the function follows a root function (Figure 3b) and the parameters of $E(V)$ can be extracted to $\Phi_B = 0.2$ and a slope of 0.1. The final function for leakage current

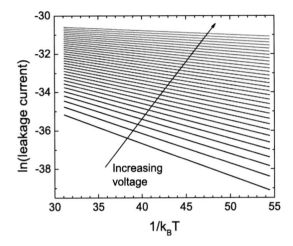

Figure 2: Arrhenius plot of the leakage current in Figure 1. The linearity of lines indicate a linear dependence of $E(V, T)$ with $1/k_B T$.

is in this example $j(V, T) = 10^{-13} \times e^{-(0.2-0.1\sqrt{V})/k_B T}$.

In the previous extraction of parameters, the pre-exponential factor $A(V, T) = A = const$ has been kept constant. This is not true for both Schottky emission $(A \propto V)$ and Poole-Frenkel conduction $(A \propto T^2)$ according to Ref. *Sze et al.* (p. 403; 2nd Ed.)[70]. Although the shape of the leakage current curves is altered, the change is not significant. Any pre-factor only contributes as a logarithmic function to the overall exponential function, resulting in a close-to-linear non-linearity of the plots mentioned earlier. Therefore the extraction of the barrier height Φ_B is comparable with a minor increase of error.

(a)

(b)

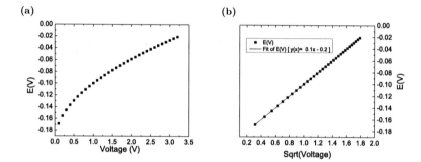

Figure 3: Activation energy plots of leakage currents in dependence of voltage. 3a) The barrier height Φ_B can be extracted for the curve at an extrapolated intercept $V = 0V$. 3b) The barrier height Φ_B is extracted from the intercept of the linear fit (red line). This plot requires information of the curvature of $E(V)$ (in this case $E(V) \propto \sqrt{V}$) to achieve a linear dependence.

Bibliography

[1] Seong Keun Kim, Sang Woon Lee, Jeong Hwan Han, Bora Lee, Seungwu Han, and Cheol Seong Hwang. Capacitors with an equivalent oxide thickness of 0.5 nm for nanoscale electronic semiconductor memory. *Advanced Functional Materials*, 20(18):2989–3003, September 2010.

[2] D. James. Recent innovations in DRAM manufacturing. In *Advanced Semiconductor Manufacturing Conference (ASMC), 2010 IEEE/SEMI*, pages 264–269. IEEE, July 2010.

[3] A. S. Bhalla, R. Guo, and R. Roy. The perovskite structure - a review of its role in ceramic science and technology. *Material Research Innovations*, 4(1):3–26, November 2000.

[4] Komatsu Y Yamanaka T, Hirai M. *Structure change of $Ca_{1-x}Sr_xTiO_3$ perovskite with composition and pressure Sample: $CaTiO_3$, x = 0.0*. Springer Press, 2002.

[5] V. V. Lemanov, A. V. Sotnikov, E. P. Smirnova, M. Weihnacht, and R. Kunze. Perovskite $CaTiO_3$ as an incipient ferroelectric. *Solid State Communications*, 110(11):611–614, 1999.

[6] Takamitsu Yamanaka, Noriyuki Hirai, and Yutaka Komatsu. Structure change of $Ca_{1-x}Sr_xTiO_3$ perovskite with composition and pressure. *American Mineralogist*, 87(8-9):1183–1189, August 2002.

[7] P. M. Woodward. Octahedral tilting in perovskites. II. structure stabilizing forces. *Acta Crystallographica Section B Structural Science*, 53(1):44–66, February 1997.

[8] K. Parlinskia, Y. Kawazoe, and Y. Waseda. Ab initio studies of phonons in $CaTiO_3$. *Journal of chemical physics*, 114(5), 2001.

[9] Alfred Kersch and Dominik Fischer. Phase stability and dielectric constant of ABO_3 perovskites from first principles. *Journal of Applied Physics*, 106(1):014105, 2009.

[10] Masatomo Yashima and Roushown Ali. Structural phase transition and octahedral tilting in the calcium titanate perovskite $CaTiO_3$. *Solid State Ionics*, 180(2-3):120–126, 2009.

[11] A. Boudali, A. Abada, M. Driss Khodja, B. Amrani, K. Amara, F. Driss Khodja, and A. Elias. Calculation of structural, elastic, electronic, and thermal properties of orthorhombic $CaTiO_3$. *Physica B: Condensed Matter*, 405(18):3879–3884, September 2010.

[12] K. Ueda, H. Yanagi, R. Noshiro, H. Hosono, and H. Kawazoe. Vacuum ultraviolet reflectance and electron energy loss spectra of. *Journal of Physics: Condensed Matter*, 10:3669–3677, 1998.

[13] K. Ueda, H. Yanagi, H. Hosono, and H. Kawazoe. Study on electronic structure of $CaTiO_3$ by spectroscopic measurements and energy band calculations. *Journal of Physics: Condensed Matter*, 11:3535, 1999.

[14] M. Arita, H. Sato, M. Higashi, K. Yoshikawa, K. Shimada, M. Nakatake, Y. Ueda, H. Namatame, M. Taniguchi, M. Tsubota, F. Iga, and T. Takabatake. Unoccupied electronic structure of $Y_{1-x}Ca_xTiO_3$ investigated by inverse photoemission spectroscopy. *Physical Review B*, 75(20):205124, 2007.

[15] A. Linz and K. Herrington. Electrical and optical properties of synthetic calcium titanate crystal. *The Journal of Chemical Physics*, 28(5):824–825, May 1958.

[16] L.S. Cavalcante, V.S. Marques, J.C. Sczancoski, M.T. Escote, M.R. Joya, J.A. Varela, M.R.M.C. Santos, P.S. Pizani, and E. Longo. Synthesis, structural refinement and optical behavior of $CaTiO_3$ powders: A comparative study of processing in different furnaces. *Chemical Engineering Journal*, 143(1-3):299–307, September 2008.

[17] Mario L. Moreira, Elaine C. Paris, Gabriela S. do Nascimento, Valeria M. Longo, Julio R. Sambrano, Valmor R. Mastelaro, Maria I.B. Bernardi, Juan Andrés, José A. Varela, and Elson Longo. Structural and optical properties of $CaTiO_3$ perovskite-based materials obtained by microwave-assisted hydrothermal synthesis: An experimental and theoretical insight. *Acta Materialia*, 57(17):5174–5185, 2009.

[18] R. P. Wang and C. J. Tao. Nb-doped $CaTiO_3$ transparent semiconductor thin films. *Journal of Crystal Growth*, 245(1-2):63–66, November 2002.

[19] S. Saha, T.P. Sinha, and A. Mookerjee. First principles study of electronic structure and optical properties of CaTiO$_3$. *The European Physical Journal B - Condensed Matter and Complex Systems*, 18(2):207–214, November 2000.

[20] R. I. Eglitis. First-principles calculations of the atomic and electronic structure of CaTiO$_3$ (111) surfaces. *Ferroelectrics*, 424(1):1–6, 2011.

[21] R I Eglitis and M Rohlfing. First-principles calculations of the atomic and electronic structure of SrZrO$_3$ and PbZrO$_3$ (001) and (011) surfaces. *Journal of Physics: Condensed Matter*, 22(41):415901, October 2010.

[22] Eric Cockayne and Benjamin P. Burton. Phonons and static dielectric constant in CaTiO$_3$ from first principles. *Physical Review B*, 62(6):3735, 2000.

[23] M. Adachi, Y. Akishige, T. Asahi, K. Deguchi, K. Gesi, K. Hasebe, T. Hikita, T. Ikeda, Y. Iwata, M. Komukae, T. Mitsui, E. Nakamura, N. Nakatani, M. Okuyama, T. Osaka, A. Sakai, E. Sawaguchi, Y. Shiozaki, T. Takenaka, K. Toyoda, T. Tsukamoto, and T. Yagi. CaTiO$_3$, 1A-7. In *Oxides*, volume 36A1, pages 1–18. Springer, Berlin, 2002.

[24] A. von Hippel. Ferroelectricity, domain structure, and phase transitions of barium titanate. *Reviews of Modern Physics*, 22(3):221, July 1950.

[25] O. Kvyatkovskii. Quantum effects in incipient and low-temperature ferroelectrics (a review). *Physics of the Solid State*, 43(8):1401–1419, 2001.

[26] W. Zhong and David Vanderbilt. Effect of quantum fluctuations on structural phase transitions in SrTiO$_3$ and BaTiO$_3$. *Physical Review B*, 53(9):5047–5050, 1996.

[27] Orest G. Vendik and Svetlana P. Zubko. Modeling the dielectric response of incipient ferroelectrics. *Journal of Applied Physics*, 82(9):4475–4483, November 1997.

[28] John H. Barrett. Dielectric constant in perovskite type crystals. *Physical Review*, 86(1):118, April 1952.

[29] V. V. Lemanov, A. V. Sotnikov, E. P. Smirnova, and M. Weihnacht. From incipient ferroelectricity in CaTiO$_3$ to real ferroelectricity in Ca$_{1-x}$Pb$_x$TiO$_3$ solid solutions. *Applied Physics Letters*, 81(5):886–888, July 2002.

[30] V.S. Marques, L.S. Cavalcante, J.C. Sczancoski, D.P. Volanti, J.W.M. Espinosa, M.R. Joya, M.R.M.C. Santos, P.S. Pizani, J.A. Varela, and E. Longo. Influence of

microwave energy on structural and photoluminescent behavior of CaTiO$_3$ powders. *Solid State Sciences*, 10(8):1056–1061, August 2008.

[31] Christopher Ashman, C. Hellberg, and Samed Halilov. Ferroelectricity in strained Ca$_{0.5}$Sr$_{0.5}$TiO$_3$ from first principles. *Physical Review B*, 82(2), July 2010.

[32] L Goncalves-Ferreira, SAT Redfern, E Artacho, and EKH Salje. Ferrielectric twin walls in CaTiO$_3$. *PHYSICAL REVIEW LETTERS*, 101(9), August 2008.

[33] E. Vlahos, T. Lummen, R. Haislmaier, S. Denev, C. Brooks, M. Biegalski, D. Schlom, C.J. Eklund, K. Rabe, C. Fennie, et al. Ferroelectricity in CaTiO$_3$ single crystal surfaces and thin films and probed by nonlinear optics and raman spectroscopy. In *APS Meeting Abstracts*, volume 1, page 22004, 2011.

[34] Sandra Van Aert, Stuart Turner, Rémi Delville, Dominique Schryvers, Gustaaf Van Tendeloo, and Ekhard K. H Salje. Direct observation of ferrielectricity at ferroelastic domain boundaries in CaTiO$_3$ by electron microscopy. *Advanced Materials*, 24(4):523–527, January 2012.

[35] YX Wang, XT Su, and WL Zhong. Electronic structure of incipient ferroelectric CaTiO$_3$. *CHINESE JOURNAL OF CHEMICAL PHYSICS*, 15(1):29–32, February 2002.

[36] C.-J. Eklund, C. J. Fennie, and K. M. Rabe. Strain-induced ferroelectricity in orthorhombic CaTiO$_3$ from first principles. *Physical Review B*, 79(22):220101, June 2009.

[37] P. Boutinaud, E. Pinel, M. Dubois, A.P. Vink, and R. Mahiou. UV-to-red relaxation pathways in CaTiO$_3$:Pr3+. *Journal of Luminescence*, 111(1-2):69–80, January 2005.

[38] Yuexiao Pan, Qiang Su, Huifang Xu, Tianhu Chen, Weikun Ge, Chunlei Yang, and Mingmei Wu. Synthesis and red luminescence of Pr3+-doped CaTiO$_3$ nanophosphor from polymer precursor. *Journal of Solid State Chemistry*, 174(1):69–73, August 2003.

[39] K. Wakino. Recent development of dielectric resonator materials and filters in japan. *Ferroelectrics*, 91:69–86, March 1989.

[40] A. Pashkin, S. Kamba, M. Berta, J. Petzelt, G. D. C. Csete de Györgyfalva, H. Zheng, H. Bagshaw, and I. M. Reaney. High frequency dielectric properties

of $CaTiO_3$-based microwave ceramics. *Journal of Physics D: Applied Physics*, 38(5):741–748, March 2005.

[41] Byoung Duk Lee, Ki Hyun Yoon, Eung Soo Kim, and Tae Hong Kim. Microwave dielectric properties of $CaTiO_3$ and $MgTiO_3$ thin films. *Japanese Journal of Applied Physics*, 42:6158–6161, 2003.

[42] Yuan-Bin Chen. Improved high Q value of $MgTiO_3$-$CaTiO_3$ microwave dielectric resonator using WO_3-doped at lower sintering temperature for microwave applications. *Journal of Alloys and Compounds*, 478(1-2):657–660, June 2009.

[43] Ailing Zhu, Jianchuan Wang, Dongdong Zhao, and Yong Du. Native defects and Pr impurities in orthorhombic $CaTiO_3$ by first-principles calculations. *Physica B: Condensed Matter*, 406(13):2697–2702, July 2011.

[44] Miguel Manso, Michel Langlet, and J. M. Martinez-Duart. Testing sol-gel $CaTiO_3$ coatings for biocompatible applications. *Materials Science and Engineering: C*, 23(3):447–450, 2003.

[45] Naota Sugiyama, HaiYan Xu, Takamasa Onoki, Yasuto Hoshikawa, Tomoaki Watanabe, Nobuhiro Matsushita, Xinmin Wang, FengXiang Qin, Mikio Fukuhara, Masahiro Tsukamoto, Nobuyuki Abe, Yuichi Komizo, Akihisa Inoue, and Masahiro Yoshimura. Bioactive titanate nanomesh layer on the ti-based bulk metallic glass by hydrothermal-electrochemical technique. *Acta Biomaterialia*, 5(4):1367–1373, 2009.

[46] Samuel Holliday and Andrei Stanishevsky. Crystallization of $CaTiO_3$ by sol-gel synthesis and rapid thermal processing. *Surface and Coatings Technology*, 188-189:741–744, 2004.

[47] John Robertson. Band offsets of wide-band-gap oxides and implications for future electronic devices. *Journal of Vacuum Science & Technology B: Microelectronics and Nanometer Structures*, 18:1785, 2000.

[48] A.J. Hartmann, M. Neilson, R.N. Lamb, K. Watanabe, and J.F. Scott. Ruthenium oxide and strontium ruthenate electrodes for ferroelectric thin-films capacitors. *Applied Physics A: Materials Science & Processing*, 70(2):239–242, February 2000.

[49] Sang Young Lee, Seong Keun Kim, Kyung Min Kim, Gyu-Jin Choi, Jeong Hwan Han, and Cheol Seong Hwang. Electrically benign Ru wet etching method for fabri-

cating Ru/TiO$_2$/Ru capacitor. *Journal of The Electrochemical Society*, 158(3):G47–G51, 2011.

[50] Dario Galizzioli, Franco Tantardini, and Sergio Trasatti. Ruthenium dioxide: a new electrode material. i. behaviour in acid solutions of inert electrolytes. *Journal of Applied Electrochemistry*, 4(1):57–67, February 1974.

[51] D. Galizzioli, F. Tantardini, and S. Trasatti. Ruthenium dioxide: a new electrode material. II. non-stoichiometry and energetics of electrode reactions in acid solutions. *Journal of Applied Electrochemistry*, 5(3):203–214, 1975.

[52] R. J. Bouchard and J. L. Gillson. Electrical properties of CaRuO$_3$ and SrRuO$_3$ single crystals. *Materials Research Bulletin*, 7(9):873–878, 1972.

[53] Wolfgang Bensch, Helmut W. Schmalle, and Armin Reller. Structure and thermochemical reactivity of CaRuO$_3$ and SrRuO$_3$. *Solid State Ionics*, 43:171–177, November 1990.

[54] Akihiko Ito, Hiroshi Masumoto, Takashi Goto, and Shunichi Sato. Characterization of alkaline earth metals ruthenate thin films. *Journal of the European Ceramic Society*, In Press, Corrected Proof, 2009.

[55] M. Darnon, T. Chevolleau, D. Eon, L. Vallier, J. Torres, and O. Joubert. Etching characteristics of TiN used as hard mask in dielectric etch process. *Journal of Vacuum Science & Technology B: Microelectronics and Nanometer Structures*, 24(5):2262, 2006.

[56] Rainer Waser. *Nanoelectronics and Information Technology: Materials, Processes, Devices*. Wiley-VCH Verlag GmbH & Co. KGaA, 2. korrigierte auflage edition, February 2005.

[57] Ch. Wenger, G. Lupina, M. Lukosius, O. Seifarth, H.-J. Müssig, S. Pasko, and Ch. Lohe. Microscopic model for the nonlinear behavior of high-k metal-insulator-metal capacitors. *Journal of Applied Physics*, 103(10):104103, 2008.

[58] S. Becu, S. Cremer, O. Noblanc, J.-L. Autran, and Delpech P. Characterization and modeling of Al$_2$O$_3$ mim capacitors: temperature and electrical field effects. In *Solid-State Device Research Conference, 2005. ESSDERC 2005. Proceedings of 35th European*, pages 265 – 268, September 2005.

[59] Stephane Becu, Sebastien Cremer, and Jean-Luc Autran. Microscopic model for dielectric constant in metal-insulator-metal capacitors with high-permittivity metallic oxides. *Applied Physics Letters*, 88(5):052902 –052902–3, January 2006.

[60] Sun Jung Kim, Byung Jin Cho, Ming-Fu Li, Shi-Jin Ding, Chunxiang Zhu, Ming Bin Yu, B. Narayanan, A. Chin, and Dim-Lee Kwong. Improvement of voltage linearity in high- kappa; MIM capacitors using HfO_2-SiO_2 stacked dielectric. *IEEE Electron Device Letters*, 25(8):538 – 540, August 2004.

[61] Jian-Jun Yang, Jing-De Chen, R. Wise, P. Steinmann, Ming-Bin Yu, Dim-Lee Kwong, Ming-Fu Li, Yee-Chia Yeo, and Chunxiang Zhu. Effective modulation of quadratic voltage coefficient of capacitance in MIM capacitors using Sm_2O_3 SiO_2 dielectric stack. *Electron Device Letters, IEEE*, 30(5):460 –462, May 2009.

[62] Thanh Hoa Phung, P. Steinmann, R. Wise, Yee-Chia Yeo, and Chunxiang Zhu. Modeling the negative quadratic VCC of SiO_2 in MIM capacitor. *IEEE Electron Device Letters*, 32(12):1671–1673, December 2011.

[63] S. Blonkowski. Nonlinear capacitance variations in amorphous oxide metal-insulator-metal structures. *Applied Physics Letters*, 91(17):172903, 2007.

[64] S. Blonkowski, E. Defay, and X. Biquard. Sign of the nonlinear dielectric susceptibility of amorphous and crystalline $SrTiO_3$ films. *Physical Review B*, 79(10):104108, 2009.

[65] F El Kamel and P Gonon. Impact of oxygen vacancy related defects on the electrical properties of $BaTiO_3$ based metal-insulator-metal devices. *IOP Conference Series: Materials Science and Engineering*, 8:012030, February 2010.

[66] F. El Kamel, P. Gonon, and C. Vallée. Experimental evidence for the role of electrodes and oxygen vacancies in voltage nonlinearities observed in high-k metal-insulator-metal capacitors. *Applied Physics Letters*, 91(17):172909–172909–3, October 2007.

[67] P. Gonon and C. Vallé. Modeling of nonlinearities in the capacitance-voltage characteristics of high-k metal-insulator-metal capacitors. *Applied Physics Letters*, 90(14):142906–142906–3, April 2007.

[68] C. Vallée, P. Gonon, C. Jorel, and F. El Kamel. Electrode oxygen-affinity influence on voltage nonlinearities in high-k metal-insulator-metal capacitors. *Applied Physics Letters*, 96(23):233504–233504–3, June 2010.

[69] Gunther Jegert, Alfred Kersch, Wenke Weinreich, Uwe Schröder, and Paolo Lugli. Modeling of leakage currents in high-k dielectrics: Three-dimensional approach via kinetic Monte Carlo. *Applied Physics Letters*, 96(6):062113, 2010.

[70] Simon M. Sze and Kwok K. Ng. *Physics of Semiconductor Devices*. John Wiley & Sons, 3. auflage edition, November 2006.

[71] S-H Lo, D.A. Buchanan, Y. Taur, and W.I. Wang. Quantum-mechanical modeling of electron tunneling current from the inversion layer of ultra-thin-oxide nMOS-FET's. *IEEE Electron Device Letters*, 18(5):209–211, 1997.

[72] John G. Simmons. Poole-frenkel effect and schottky effect in metal-insulator-metal systems. *Physical Review*, 155(3):657–660, 1967.

[73] Doo Seok Jeong and Cheol Seong Hwang. Tunneling-assisted poole-frenkel conduction mechanism in HfO$_2$ thin films. *Journal of Applied Physics*, 98:113701, 2005.

[74] K. Xiong, J. Robertson, M. C. Gibson, and S. J. Clark. Defect energy levels in HfO$_2$ high-dielectric-constant gate oxide. *Applied Physics Letters*, 87(18):183505, 2005.

[75] J. L. Gavartin, D. Munoz Ramo, A. L. Shluger, G. Bersuker, and B. H. Lee. Negative oxygen vacancies in HfO$_2$ as charge traps in high-k stacks. *Applied Physics Letters*, 89(8):082908, 2006.

[76] Hyoungsub Kim, Ann Marshall, Paul C. McIntyre, and Krishna C. Saraswat. Crystallization kinetics and microstructure-dependent leakage current behavior of ultra-thin HfO$_2$ dielectrics: In situ annealing studies. *Applied Physics Letters*, 84:2064, 2004.

[77] M. Houssa, M. Tuominen, M. Naili, V. Afanasev, A. Stesmans, S. Haukka, and M. M Heyns. Trap-assisted tunneling in high permittivity gate dielectric stacks. *Journal of Applied Physics*, 87(12):8615–8620, June 2000.

[78] A. Gushterov and S. Simeonov. Trap-assisted tunnelling in ion-implanted n-Si/SiO$_2$ structures. *Vacuum*, 76(2-3):315–318, November 2004.

[79] A. Gushterov and S. Simeonov. Extraction of trap-assisted tunneling parameters by graphical method in thin n-Si/SiO$_2$ structures. *Journal of Optoelectronics and Advanced Materials*, 7(3):1389–1393, 2005.

[80] M. Schumacher, S. Manetta, and R. Waser. Dielectric relaxation phenomena in superparaelectric and ferroelectric ceramic thin films and the relevance with respect to high density DRAM and FRAM applications. *Le Journal de Physique IV*, 08(PR9):4, 1998.

[81] Andrew K Jonscher. Dielectric relaxation in solids. *Journal of Physics D: Applied Physics*, 32:R57–R70, July 1999.

[82] M. Schumacher, G. W. Dietz, and R. Waser. Dielectric relaxation of perovskite-type oxide thin films. *Integrated Ferroelectrics*, 10:231–245, October 1995.

[83] Noboru Mikami. (Ba,Sr)TiO$_3$ films and process integration for Dram capacitor. In R. Ramesh, editor, *Thin Film Ferroelectric Materials and Devices*, volume 3 of *Electronic Materials: Science & Technology*, pages 43–70. Springer US, 1997.

[84] Siegfried Hunklinger. *Festkörperphysik*. Oldenbourg Verlag, 2007.

[85] Vasil Yanev, Mathias Rommel, Martin Lemberger, Silke Petersen, Brigitte Amon, Tobias Erlbacher, Anton J. Bauer, Heiner Ryssel, Albena Paskaleva, Wenke Weinreich, Christian Fachmann, Johannes Heitmann, and Uwe Schroeder. Tunneling atomic-force microscopy as a highly sensitive mapping tool for the characterization of film morphology in thin high-k dielectrics. *Applied Physics Letters*, 92(25):252910, 2008.

[86] O. Bierwagen, L. Geelhaar, X. Gay, M. Piesins, H. Riechert, B. Jobst, and A. Rucki. Leakage currents at crystallites in ZrAl$_x$O$_y$ thin films measured by conductive atomic-force microscopy. *Applied Physics Letters*, 90(23):232901, 2007.

[87] Dominik Martin, Matthias Grube, Walter M. Weber, Jürgen Rüstig, Oliver Bierwagen, Lutz Geelhaar, and Henning Riechert. Local charge transport in nanoscale amorphous and crystalline regions of high-k (ZrO$_2$)$_{0.8}$(Al$_2$O$_3$)$_{0.2}$ thin films. *Applied Physics Letters*, 95(14):142906, 2009.

[88] Tomonori Aoyama, Shigehiko Saida, Yasunori Okayama, Masanori Fujisaki, Keitaro Imai, and Tsunetoshi Arikado. Leakage current mechanism of amorphous and polycrystalline Ta$_2$O$_5$ films grown by chemical vapor deposition. *Journal of The Electrochemical Society*, 143(3):977–983, 1996.

[89] Fu-Chien Chiu, Jenn-Jyh Wang, Joseph Ya-min Lee, and Shich Chuan Wu. Leakage currents in amorphous Ta$_2$O$_5$ thin films. *Journal of Applied Physics*, 81:6911, 1997.

135

[90] H. M. Gupta and R. J. Van Overstraeten. Role of trap states in the insulator region for MIM characteristics. *Journal of Applied Physics*, 46:2675, 1975.

[91] Gerhard Franz. *Niederdruckplasmen und Mikrostrukturtechnik*. Springer Berlin Heidelberg, 3. aufl. edition, December 2003.

[92] Bestec GmbH. *Sputter Cluster Reference Manual*. Am Studio 2b, D-12489 Berlin, 2008.

[93] David Briggs and John. T. Grant. *Surface Analysis by Auger and X-Ray Photoelectron Spectroscopy*. IM Publications, 2003.

[94] G.M. Ertl and J. Küppers. *Low energy electrons and surface chemistry*. VCH-Verlag, Weinheim, 1985.

[95] Stefan Hüfner. *Photoelectron Spectroscopy: Principles and Applications*. Springer, 1995.

[96] M. Cardona and L. Ley. *Photoemission in Solids I: General Principles*, volume 26. Springer-Verlag, Berlin, 1978.

[97] H. Ibach. *Electron Spectroscopy for Surface Analysis*. Springer-Verlag, Berlin, 1977.

[98] R F Egerton. Electron energy-loss spectroscopy in the TEM. *Reports on Progress in Physics*, 72(1):016502, January 2009.

[99] R. F. Egerton. *Electron Energy-Loss Spectroscopy in the Electron Microscope*. Springer, July 2011.

[100] Hans Lüth. *Solid Surfaces, Interfaces and Thin Films*. Springer, September 2010.

[101] Dieter K. Schroder. *Semiconductor Material and Device Characterization*. John Wiley & Sons, 3. auflage edition, February 2006.

[102] David B. Williams and C. Barry Carter. *Transmission Electron Microscopy: A Textbook for Materials Science*. Springer US, 2nd ed. edition, August 2009.

[103] Mark Calleja, Martin T Dove, and Ekhard K H Salje. Trapping of oxygen vacancies on twin walls of $CaTiO_3$: a computer simulation study. *Journal of Physics: Condensed Matter*, 15(14):2301–2307, April 2003.

[104] J G Simmons. Conduction in thin dielectric films. *Journal of Physics D: Applied Physics*, 4(5):613–657, May 1971.

[105] J. Robertson and C. W. Chen. Schottky barrier heights of tantalum oxide, barium strontium titanate, lead titanate, and strontium bismuth tantalate. *Applied Physics Letters*, 74:1168, 1999.

[106] Keithley. *Model 4200-SCS Semiconductor Characterization System Reference Manual*, rev. 1 edition, May 2010.

[107] C. Zhou and D. M. Newns. Intrinsic dead layer effect and the performance of ferroelectric thin film capacitors. *Journal of Applied Physics*, 82(6):3081, 1997.

[108] A. M. Bratkovsky and A. P. Levanyuk. Very large dielectric response of thin ferroelectric films with the dead layers. *Physical Review B*, 63(13):132103, 2001.

[109] Massimiliano Stengel and Nicola A. Spaldin. Origin of the dielectric dead layer in nanoscale capacitors. *Nature*, 443(7112):679–682, 2006.

[110] Samuel Ruben. *Handbook of the Elements*. Open Court Publishing Company, 1985.

[111] W.J. Garceau, P.R. Fournier, and G.K. Herb. TiN as a diffusion barrier in the Ti-Pt-Au beam-lead metal system. *Thin Solid Films*, 60(2):237–247, June 1979.

[112] I. Montero, C. Jiménez, and J. Perrière. Surface oxidation of TiN_x films. *Surface Science*, 251-252(0):1038–1043, July 1991.

[113] W. Weinreich, R. Reiche, M. Lemberger, G. Jegert, J. Müller, L. Wilde, S. Teichert, J. Heitmann, E. Erben, L. Oberbeck, U. Schröder, A.J. Bauer, and H. Ryssel. Impact of interface variations on J-V and C-V polarity asymmetry of MIM capacitors with amorphous and crystalline $Zr_{1-x}Al_xO_2$ films. *Microelectronic Engineering*, 86(7-9):1826–1829, September 2009.

[114] S. Bhaskar, P. S. Dobal, S. B. Majumder, and R. S. Katiyar. X-ray photoelectron spectroscopy and micro-raman analysis of conductive RuO_2 thin films. *Journal of Applied Physics*, 89:2987, 2001.

[115] C. S. Petersson, J. E. E. Baglin, J. J. Dempsey, F. M. d'Heurle, and S. J. La Placa. Silicides of ruthenium and osmium: Thin film reactions, diffusion, nucleation, and stability. *Journal of Applied Physics*, 53(7):4866–4883, July 1982.

[116] D. Donoval, L. Stolt, H. Norde, J. de Sousa Pires, P. A. Tove, and C. S. Petersson. Barrier heights to silicon, of ruthenium and its silicide. *Journal of Applied Physics*, 53(7):5352–5353, July 1982.

[117] H.-C. Wen, P. Lysaght, H. N. Alshareef, C. Huffman, H. R. Harris, K. Choi, Y. Senzaki, H. Luan, P. Majhi, B. H. Lee, M. J. Campin, B. Foran, G. D. Lian, and D.-L. Kwong. Thermal response of Ru electrodes in contact with SiO_2 and Hf-based high-k gate dielectrics. *Journal of Applied Physics*, 98(4):043520, 2005.

[118] Silze F. *Präparation nanoskaliger temperaturstabiler Pt-Elektroden mit RuO_2-Diffusionsbarriere für Kondensatoren in DRAM*. PhD thesis, Institut für Werkstoffwissenschaft, Faculty of Mechanical Science and Engineering, TU Dresden, 2011.

[119] F. Kreupl, R. Bruchhaus, P. Majewski, J.B. Philipp, R. Symanczyk, T. Happ, C. Arndt, M. Vogt, R. Zimmermann, A. Buerke, A.P. Graham, and M. Kund. Carbon-based resistive memory. In *Electron Devices Meeting, 2008. IEDM 2008. IEEE International*, pages 1 –4, December 2008.

[120] G. Aichmayr, A. Avellan, G.S. Duesberg, F. Kreupl, S. Kudelka, and M. Liebau. Carbon / high-k trench capacitor for the 40nm DRAM generation. In *2007 IEEE Symposium on VLSI Technology*, pages 186 –187, June 2007.

[121] A. P Graham, G. Schindler, G. S Duesberg, T. Lutz, and W. Weber. An investigation of the electrical properties of pyrolytic carbon in reduced dimensions: Vias and wires. *Journal of Applied Physics*, 107(11):114316–114316-4, June 2010.

[122] A. P. Graham, K. Richter, T. Jay, W. Weber, S. Knebel, U. Schröder, and T. Mikolajick. An investigation of the electrical properties of metal-insulator-silicon capacitors with pyrolytic carbon electrodes. *Journal of Applied Physics*, 108(10):104508, 2010.

[123] H. Haneda, I. Sakaguchi, S. Hishita, T. Ishigaki, and T. Mitsuhashi. Oxygen metastable defects in calcium titanate thin films. *Journal of Thermal Analysis and Calorimetry*, 60(2):675–681, 2000.

[124] RH Deitch, EJ West, TG Giallorenzi, and JF Weller. Sputtered thin films for integrated optics. *Applied Optics*, 13(4):712–715, 1974.

[125] L Sarakha, A Bousquet, E Tomasella, P Boutinaud, and R Mahiou. Investigation of $CaTiO_3$:Pr3+ thin films deposited by radiofrequency reactive magnetron sputtering for electroluminescence application. *IOP Conference Series: Materials Science and Engineering*, 12:012008, June 2010.

[126] K van Benthem, R.H French, W Sigle, C Elsässer, and M Rühle. Valence electron energy loss study of Fe-doped SrTiO₃ and a Σ13 boundary: electronic structure and dispersion forces. *Ultramicroscopy*, 86(3-4):303–318, February 2001.

[127] B. Rafferty and L. M. Brown. Direct and indirect transitions in the region of the band gap using electron-energy-loss spectroscopy. *Physical Review B*, 58(16):10326–10337, 1998.

[128] S. Lazar, G.A. Botton, M.-Y. Wu, F.D. Tichelaar, and H.W. Zandbergen. Materials science applications of HREELS in near edge structure analysis and low-energy loss spectroscopy. *Ultramicroscopy*, 96(3-4):535–546, September 2003.

[129] Koji Kimoto, Gerald Kothleitner, Werner Grogger, Yoshio Matsui, and Ferdinand Hofer. Advantages of a monochromator for bandgap measurements using electron energy-loss spectroscopy. *Micron*, 36(2):185–189, February 2005.

[130] K. van Benthem, C. Elsässer, and R. H French. Bulk electronic structure of SrTiO₃ experiment and theory. *Journal of Applied Physics*, 90(12):6156–6164, December 2001.

[131] J. Padilla and David Vanderbilt. Ab initio study of SrTiO₃ surfaces. *Surface Science*, 418(1):64–70, November 1998.

[132] L. Névot and P. Croce. Caractérisation des surfaces par réflexion rasante de rayons x. application à l'étude du polissage de quelques verres silicates. *Revue de Physique Appliquée*, 15(3):761–779, 1980.

[133] Michihiro Murata, Kikuo Wakino, and Shigero Ikeda. X-ray photoelectron spectroscopic study of perovskite titanates and related compounds: An example of the effect of polarization on chemical shifts. *Journal of Electron Spectroscopy and Related Phenomena*, 6(5):459–464, 1975.

[134] T. Hanawa, H. Ukai, and K. Murakami. X-ray photoelectron spectroscopy of calcium-ion-implanted titanium. *Journal of Electron Spectroscopy and Related Phenomena*, 63(4):347–354, November 1993.

[135] B. Demri and D. Muster. XPS study of some calcium compounds. *Journal of Materials Processing Technology*, 55(3-4):311–314, 1995.

[136] K. Asami, N. Ohtsu, K. Saito, and T. Hanawa. CaTiO₃ films sputter-deposited under simultaneous Ti-ion implantation on Ti-substrate. *Surface and Coatings Technology*, 200(1-4):1005–1008, 2005.

[137] Th. Proffen, S. J. L. Billinge, T. Egami, and D. Louca. Structural analysis of complex materials using the atomic pair distribution function - a practical guide. *Zeitschrift für Kristallographie*, 218(2-2003):132–143, February 2003.

[138] G. Dalba, P. Fornasini, M. Grazioli, and F. Rocca. Local disorder in crystalline and amorphous germanium. *Physical Review B*, 52(15):11034–11043, 1995.

[139] P. H. Fuoss, P. Eisenberger, W. K. Warburton, and A. Bienenstock. Application of differential anomalous x-ray scattering to structural studies of amorphous materials. *Physical Review Letters*, 46(23):1537–1540, June 1981.

[140] Satoshi Hashimoto and Akihiro Tanaka. Alteration of Ti 2p XPS spectrum for titanium oxide by low-energy Ar ion bombardment. *Surface and Interface Analysis*, 34(1):262–265, 2002.

[141] Hirohide Nakamatsu, Hirohiko Adachi, and Shigero Ikeda. Electronic structure of the valence band for perovskite-type titanium double oxides studied by XPS and DV-X[alpha] cluster calculations. *Journal of Electron Spectroscopy and Related Phenomena*, 24(2):149–159, 1981.

[142] J. F. Moulder, W. F. Stickle, P. E. Sobol, and K. D. Bomben. *Handbook of X-ray Photoelectron Spectroscopy*. Perkin-Elmer Cooperation, 1992.

[143] S. Kačiulis, G. Mattogno, L. Pandolfi, M. Cavalli, G. Gnappi, and A. Montenero. XPS study of apatite-based coatings prepared by sol-gel technique. *Applied Surface Science*, 151(1-2):1–5, September 1999.

[144] D.-J. Won, C.-H. Wang, H.-K. Jang, and D.-J. Choi. Effects of thermally induced anatase-to-rutile phase transition in MOCVD-grown TiO_2 films on structural and optical properties. *Applied Physics A: Materials Science & Processing*, 73(5):595–600, 2001.

[145] M. Grube, D. Martin, W.M. Weber, T. Mikolajick, and H. Riechert. Phase stabilization of sputtered strontium zirconate. *Microelectronic Engineering*, 88(7):1326–1329, July 2011.

[146] N. Menou, X. P Wang, B. Kaczer, W. Polspoel, M. Popovici, K. Opsomer, M. A Pawlak, W. Knaepen, C. Detavernier, T. Blomberg, D. Pierreux, J. Swerts, J. W Maes, P. Favia, H. Bender, B. Brijs, W. Vandervorst, S. Van Elshocht, D. J Wouters, S. Biesemans, and J. A Kittl. 0.5 nm EOT low leakage ALD $SrTiO_3$

on TiN MIM capacitors for DRAM applications. In *Electron Devices Meeting, 2008. IEDM 2008. IEEE International*, pages 1–4. IEEE, December 2008.

[147] Shahin A Mojarad, Kelvin S. K Kwa, Jonathan P Goss, Zhiyong Zhou, Nikhil K Ponon, Daniel J. R Appleby, Raied A. S Al-Hamadany, and Anthony O'Neill. A comprehensive study on the leakage current mechanisms of $Pt/SrTiO_3/Pt$ capacitor. *Journal of Applied Physics*, 111(1):014503–014503–6, January 2012.

[148] A.P.M. Kentgens, A.H. Carim, and B. Dam. Transmission electron microscopy of thin $YBa_2Cu_3O_{7-x}$ films on (001) $SrTiO_3$ prepared by DC triode sputtering. *Journal of Crystal Growth*, 91(3):355–362, August 1988.

[149] S. R. Singh and J. M. Howe. Studies on the deformation behaviour of interfaces in $(\gamma + \alpha 2)$ titanium aluminide by high-resolution transmission electron microscopy. *Philosophical Magazine Letters*, 65(5):233–241, 1992.

[150] F. A. Ponce, B. S. Krusor, J. S. Major, W. E. Plano, and D. F. Welch. Microstructure of GaN epitaxy on SiC using AlN buffer layers. *Applied Physics Letters*, 67(3):410–412, July 1995.

[151] T. M. Smeeton, M. J. Kappers, J. S. Barnard, M. E. Vickers, and C. J. Humphreys. Electron-beam-induced strain within InGaN quantum wells: False indium 'cluster' detection in the transmission electron microscope. *Applied Physics Letters*, 83(26):5419–5421, December 2003.

[152] Hui Gu and Miran Ceh. Indirect EELS imaging reaching atomic scale - CaO planar faults in $CaTiO_3$. *Ultramicroscopy*, 78(1-4):221–231, June 1999.

[153] Naofumi Ohtsu, Akihiko Ito, Kesami Saito, and Takao Hanawa. Characterization of calcium titanate thin films deposited on titanium with reactive sputtering and pulsed laser depositions. *Surface and Coatings Technology*, 201(18):7686–7691, June 2007.

[154] Naofumi Ohtsu, Chikage Abe, Tetsuya Ashino, Satoshi Semboshi, and Kazuaki Wagatsuma. Calcium-hydroxide slurry processing for bioactive calcium-titanate coating on titanium. *Surface and Coatings Technology*, 202(21):5110–5115, July 2008.

[155] M. Zacharias, J. Blaïĺsing, P. Veit, L. Tsybeskov, K. Hirschman, and P. M. Fauchet. Thermal crystallization of amorphous Si/SiO_2 superlattices. *Applied Physics Letters*, 74(18):2614, 1999.

[156] M. Zacharias and P. Streitenberger. Crystallization of amorphous superlattices in the limit of ultrathin films with oxide interfaces. *Physical Review B*, 62(12):8391, 2000.

[157] Xiaoqian Wei, Shi Luping, Chong Tow Chong, Zhao Rong, and Lee Hock Koon. Thickness dependent nano-crystallization in $Ge_2Sb_2Te_5$ films and its effect on devices. *Japanese Journal of Applied Physics*, 46(4B):2211–2214, April 2007.

[158] G. R. Strobl, M. J. Schneider, and I. G. Voigt-Martin. Model of partial crystallization and melting derived from small-angle x-ray scattering and electron microscopic studies on low-density polyethylene. *Journal of Polymer Science: Polymer Physics Edition*, 18(6):1361–1381, 1980.

[159] A. Dunlop, G. Jaskierowicz, G. Rizza, and M. Kopcewicz. Partial crystallization of an amorphous alloy by electronic energy deposition. *Physical Review Letters*, 90(1):015503, January 2003.

[160] Junwoo Son, Joël Cagnon, Damien S. Boesch, and Susanne Stemmer. Epitaxial $SrTiO_3$ tunnel barriers on Pt/MgO substrates. *Applied Physics Express*, 1:061603, June 2008.

[161] Aravind Asthagiri, Christoph Niederberger, Andrew J. Francis, Lisa M. Porter, Paul A. Salvador, and David S. Sholl. Thin Pt films on the polar $SrTiO_3(1\ 1\ 1)$ surface: an experimental and theoretical study. *Surface Science*, 537(1-3):134–152, July 2003.

[162] Kazuhide Abe and Shuichi Komatsu. Epitaxial growth of $SrTiO_3$ films on pt electrodes and their electrical properties. *Japanese Journal of Applied Physics*, 31(Part 1, No. 9B):2985–2988, 1992.

[163] Anderson Janotti, Bharat Jalan, Susanne Stemmer, and Chris G. Van de Walle. Effects of doping on the lattice parameter of $SrTiO_3$. *Applied Physics Letters*, 100(26):262104, 2012.

[164] J. H. Haeni, P. Irvin, W. Chang, R. Uecker, P. Reiche, Y. L. Li, S. Choudhury, W. Tian, M. E. Hawley, B. Craigo, A. K. Tagantsev, X. Q. Pan, S. K. Streiffer, L. Q. Chen, S. W. Kirchoefer, J. Levy, and D. G. Schlom. Room-temperature ferroelectricity in strained $SrTiO_3$. *Nature*, 430:758–761, August 2004.

142

[165] P. Besson, J.P. Poirier, and G.D. Price. Dislocations in $CaTiO_3$ perovskite deformed at high-temperature: a transmission electron microscopy study. *Physics and Chemistry of Minerals*, 23(6):337–344, 1996.

[166] G. R. Harp, R. F. C. Farrow, R. F. Marks, and J. E. Vazquez. Epitaxial growth and homoepitaxy of Pt(110) and Cu(110) on $SrTiO_3$(110). *Journal of Crystal Growth*, 127(1-4):627–633, February 1993.

[167] B. M. Lairson, M. R. Visokay, R. Sinclair, S. Hagstrom, and B. M. Clemens. Epitaxial Pt(001), Pt(110), and Pt(111) films on MgO(001), MgO(110), MgO(111), and Al_2O_3(0001). *Applied Physics Letters*, 61(12):1390–1392, September 1992.

[168] J. Rankin, J. C. McCallum, and L. A. Boatner. Annealing environment effects in the epitaxial regrowth of ion-beam-amorphized layers on $CaTiO_3$. *Journal of Applied Physics*, 78(3):1519–1527, August 1995.

[169] A. Asthagiri and D.S. Sholl. DFT study of pt adsorption on low index $SrTiO_3$ surfaces: $SrTiO_3$ (1 0 0), $SrTiO_3$ (1 1 1) and $SrTiO_3$ (1 1 0). *Surface science*, 581(1):66–87, 2005.

[170] R. I. Eglitis and David Vanderbilt. Ab initio calculations of the atomic and electronic structure of $CaTiO_3$ (001) and (011) surfaces. *Physical Review B*, 78(15):155420, 2008.

[171] Jian-Min Zhang, Jie Cui, Ke-Wei Xu, Vincent Ji, and Zhen-Yong Man. Ab initio modeling of $CaTiO_3$ (110) polar surfaces. *Physical Review B*, 76(11):115426, 2007.

[172] Yuan Xu Wang, Masao Arai, Taizo Sasaki, and Chun Lei Wang. First-principles study of the (001) surface of cubic $CaTiO_3$. *Physical Review B*, 73(3):035411, January 2006.

[173] Wei Liu, Chuncheng Wang, Jie Cui, and Zhen-Yong Man. Ab initio calculations of the $CaTiO_3$ (111) polar surfaces. *Solid State Communications*, 149(43-44):1871–1876, November 2009.

[174] Andrew J. Francis and Paul A. Salvador. Chirally oriented heteroepitaxial thin films grown by pulsed laser deposition: Pt(621) on $SrTiO_3$(621). *Journal of Applied Physics*, 96(5):2482–2493, September 2004.

[175] A. Krause, W. M. Weber, U. Schröder, D. Pohl, B. Rellinghaus, J. Heitmann, and T. Mikolajick. Reduction of leakage currents with nanocrystals embedded in

an amorphous matrix in metal-insulator-metal capacitor stacks. *Applied Physics Letters*, 99(22):222905–222905–3, November 2011.

[176] Gunther Jegert, Alfred Kersch, Wenke Weinreich, and Paolo Lugli. Ultimate scaling of $TiN/ZrO_2/TiN$ capacitors: Leakage currents and limitations due to electrode roughness. *Journal of Applied Physics*, 109(1):014504–014504–6, January 2011.

[177] Jin-Seong Kim, Kyung-Hoon Cho, Lee-Seung Kang, Jong-Woo Sun, Dong-Soo Paik, Tae-Geun Seong, Chong-Yun Kang, Jong-Hee Kim, Tae-Hyun Sung, and Sahn Nahm. Microstructure and electrical properties of amorphous films grown on Cu/Ti/ /Si substrates using RF magnetron sputtering. *IEEE Transactions on Electron Devices*, 58(5):1462 –1467, May 2011.

[178] M. F. Zhou, T. Bak, J. Nowotny, M. Rekas, C. C. Sorrell, and E. R. Vance. Defect chemistry and semiconducting properties of calcium titanate. *Journal of Materials Science: Materials in Electronics*, 13(12):697–704, 2002.

[179] K. Tse and J. Robertson. Defects and their passivation in high k gate oxides. *Microelectronic Engineering*, 84(4):663–668, April 2007.

[180] L. Goux, H. Vander Meeren, and D. J. Wouters. Metallorganic chemical vapor deposition of sr-ta-o and bi-ta-o films for backend integration of high-k capacitors. *Journal of The Electrochemical Society*, 153(7):F132–F136, 2006.

[181] S. Bécu, S. Crémer, and J.L. Autran. Capacitance non-linearity study in Al_2O_3 MIM capacitors using an ionic polarization model. *Microelectronic Engineering*, 83(11-12):2422–2426, November 2006.

[182] Cem Basceri, S. K. Streiffer, Angus I. Kingon, and R. Waser. The dielectric response as a function of temperature and film thickness of fiber-textured $(Ba,Sr)TiO_3$ thin films grown by chemical vapor deposition. *Journal of Applied Physics*, 82(5):2497, 1997.

[183] S. Schmelzer, D. Bräuhaus, S. Hoffmann-Eifert, P. Meuffels, U. Böttger, L. Oberbeck, P. Reinig, U. Schröder, and R. Waser. $SrTiO_3$ thin film capacitors on silicon substrates with insignificant interfacial passive layers. *Applied Physics Letters*, 97(13):132907, 2010.

[184] ITRS. International Technology Roadmap for Semiconductors (ITRS), DRAM, 2012.

[185] K. C Chiang, Ching-Chien Huang, G. L Chen, Wen Jauh Chen, H. L Kao, Yung-Hsien Wu, A. Chin, and S. P McAlister. High-performance $SrTiO_3$ MIM capacitors for analog applications. *IEEE Transactions on Electron Devices*, 53(9):2312–2319, September 2006.

[186] M. Lukosius, T. Blomberg, D. Walczyk, G. Ruhl, and Ch Wenger. Metal-insulator-metal capacitors with ALD grown $SrTiO_3$ influence of pt electrodes. *IOP Conference Series: Materials Science and Engineering*, 41(1):012015, December 2012.

[187] C. Jorel, C. Vallée, P. Gonon, E. Gourvest, C. Dubarry, and E. Defay. High performance metal-insulator-metal capacitor using a $SrTiO_3/ZrO_2$ bilayer. *Applied Physics Letters*, 94(25):253502, June 2009.

[188] J. C. Slater. The lorentz correction in barium titanate. *Physical Review*, 78(6):748–761, June 1950.

[189] Jeong-Ho Sohn, Yoshiyuki Inaguma, Mitsuru Itoh, and Tetsuro Nakamura. Cooperative interaction of oxygen octahedra for dielectric properties in the perovskite-related layered compounds $Sr_{n+1}Ti_nO_{3n+1}$,$Ca_{n+1}Ti_nO_{3n+1}$ and $Sr_{n+1}(Ti_{0.5}Sn_{0.5})_nO_{3n+1}$ (n = 1, 2, 3 and ∞). *Materials Science and Engineering: B*, 41(1):50–54, 1996.

[190] Cristina E. Ciomaga, Maria T. Buscaglia, Vincenzo Buscaglia, and Liliana Mitoseriu. Oxygen deficiency and grain boundary-related giant relaxation in $Ba(Zr,Ti)O_3$ ceramics. *Journal of Applied Physics*, 110(11):114110–114110–7, December 2011.

[191] H. Hu, S. J. Ding, H. F. Lim, C. Zhu, M. F. Li, S. J. Kim, X. F. Yu, J. H. Chen, Y. F. Yong, and B. J. Cho. High performance ALD HfO_2-Al_2O_3 laminate MIM capacitors for RF and mixed signal IC applications. *IEDM Tech. Dig*, 15:1–15, 2003.

[192] Palani Balaya, Janez Jamnik, Jürgen Fleig, and Joachim Maier. Mesoscopic hole conduction in nanocrystalline $SrTiO_3$ a detailed analysis by impedance spectroscopy. *Journal of The Electrochemical Society*, 154(6):P69–P76, June 2007.

[193] J. Ross Macdonald. Note on the parameterization of the constant-phase admittance element. *Solid State Ionics*, 13(2):147–149, 1984.

[194] John Robertson. High dielectric constant gate oxides for metal oxide Si transistors. *Reports on Progress in Physics*, 69(2):327–396, February 2006.

[195] Yoshio Abe, Midori Kawamura, Hideto Yanagisawa, and Katsutaka Sasaki. Surface oxidation behavior of TiN film caused by depositing $SrTiO_3$ film. *Japanese Journal of Applied Physics*, 34(Part 2, No. 12B):L1678–L1681, 1995.

[196] R. D. Vispute, J. Narayan, K. Dovidenko, K. Jagannadham, N. Parikh, A. Suvkhanov, and J. D. Budai. Heteroepitaxial structures of $SrTiO_3$/TiN on si(100) by in situ pulsed laser deposition, 1996.

[197] B Hudec, K Husekova, E Dobrocka, T Lalinsky, J Aarik, A Aidla, and K Frohlich. High-permittivity metal-insulator-metal capacitors with TiO_2 rutile dielectric and RuO_2 bottom electrode. *IOP Conference Series: Materials Science and Engineering*, 8:012024, February 2010.

[198] A. Paskaleva, M. Lemberger, E. Atanassova, and A. J. Bauer. Traps and trapping phenomena and their implications on electrical behavior of high-k capacitor stacks. *Journal of Vacuum Science & Technology B: Microelectronics and Nanometer Structures*, 29(1):01AA03, 2011.

[199] Nicolas Gaillard, Luc Pinzelli, Mickael Gros-Jean, and Ahmad Bsiesy. In situ electric field simulation in metal/insulator/metal capacitors. *Applied Physics Letters*, 89(13):133506, September 2006.

[200] T. Remmel, R. Ramprasad, and J. Walls. Leakage behavior and reliability assessment of tantalum oxide dielectric MIM capacitors. In *Reliability Physics Symposium Proceedings, 2003. 41st Annual. 2003 IEEE International*, pages 277– 281. IEEE, April 2003.

[201] M. S Kim, B. Kaczer, S. Starschich, M. Popovici, J. Swerts, O. Richard, K. Tomida, C. Vrancken, S. Van Elshocht, I. Debusschere, L. Altimime, and J.A. Kittl. Understanding of Trap-Assisted Tunneling Current - Assisted by Oxygen Vacancies in $RuOx$/$SrTiO_3$/TiN MIM capacitor for the DRAM application. In *Memory Workshop (IMW), 2012 4th IEEE International*, pages 1–4, 2012.

[202] M. A. Pawlak, B. Kaczer, M.-S. Kim, M. Popovici, J. Swerts, W.-C. Wang, K. Opsomer, P. Favia, K. Tomida, A. Belmonte, B. Govoreanu, C. Vrancken, C. Demeurisse, H. Bender, V. V. Afanasév, I. Debusschere, L. Altimime, and J. A. Kittl. Impact of bottom electrode and $Sr_xTi_yO_z$ film formation on physical and electrical properties of metal-insulator-metal capacitors. *Applied Physics Letters*, 98(18):182902, 2011.

[203] S. M. Sze. Current transport and maximum dielectric strength of silicon nitride films. *Journal of Applied Physics*, 38(7):2951–2956, June 1967.

[204] A. K. Sinha and T. E. Smith. Electrical properties of Si-N films deposited on silicon from reactive plasma. *Journal of Applied Physics*, 49(5):2756–2760, May 1978.

[205] M. C. Chen, D. V. Lang, W. C. DautremontâĂŘSmith, A. M. Sergent, and J. P. Harbison. Effects of leakage current on deep level transient spectroscopy. *Applied Physics Letters*, 44(8):790–792, April 1984.

[206] Shigeaki Zaima, Takeshi Furuta, Yasuo Koide, Yukio Yasuda, and Makio Iida. Conduction mechanism of leakage current in Ta_2O_5 films on si prepared by LPCVD. *Journal of The Electrochemical Society*, 137(9):2876–2879, September 1990.

[207] Robert M. Hill. Poole-frenkel conduction in amorphous solids. *Philosophical Magazine*, 23(181):59–86, 1971.

[208] V. I. Koldyaev. Nonlinear frenkel and poole effects. *Philosophical Magazine Part B*, 79(2):331–342, 1999.

[209] K.-H. Allers. Prediction of dielectric reliability from i-v characteristics: Poole-frenkel conduction mechanism leading to $\sqrt{(e)}$ model for silicon nitride MIM capacitor. *Microelectronics Reliability*, 44(3):411–423, 2004.

[210] J. D. Baniecki, R. B. Laibowitz, T. M. Shaw, P. R. Duncombe, D. A. Neumayer, D. E. Kotecki, H. Shen, and Q. Y. Ma. Dielectric relaxation of $Ba_{0.7}Sr_{0.3}TiO_3$ thin films from 1 mHz to 20 GHz. *Applied Physics Letters*, 72(4):498–500, January 1998.

[211] J. Heitmann, F. Müller, M. Zacharias, and U. Gösele. Silicon nanocrystals: Size matters. *Advanced Materials*, 17(7):795–803, April 2005.

[212] Lin Gu, Vesna Srot, Wilfried Sigle, Christoph Koch, Peter van Aken, Ferdinand Scholz, Sarad B. Thapa, Christoph Kirchner, Michael Jetter, and Manfred Rühle. Band-gap measurements of direct and indirect semiconductors using monochromated electrons. *Physical Review B*, 75(19):195214, 2007.

[213] G.E. Jellison Jr., V.I. Merkulov, A.A. Puretzky, D.B. Geohegan, G. Eres, D.H. Lowndes, and J.B. Caughman. Characterization of thin-film amorphous semiconductors using spectroscopic ellipsometry. *Thin Solid Films*, 377-378(0):68–73, 2000.

[214] Kurt Oughstun and Natalie Cartwright. On the lorentz-lorenz formula and the lorentz model of dielectric dispersion: addendum. *Optics Express*, 11(21):2791–2792, 2003.

[215] Bernhard von Blanckenhagen, Diana Tonova, and Jens Ullmann. Application of the tauc-lorentz formulation to the interband absorption of optical coating materials. *Applied Optics*, 41(16):3137–3141, June 2002.

Acknowledgments

I want to thank all the people, who helped during my PhD time for their support regarding professional and moral help. This thesis would not be possible without them. First of all, I want to thank Prof. Dr. Thomas Mikolajick, who made it possible to write the thesis at Namlab from the beginning. His support from the scientific view was indispensable and motivated my research concerning electrical characterization. This includes the beneficial comments regarding the corrections of my thesis. Prof. Dr. Franz Kreupl, I am very thankful for correcting my thesis as well.

My supervisor Dr. Walter Weber, all progress in this work is honored to him. Although this topic of dielectrics is not his main research interest, he was the best supervisor to have. Thanks for your all-time support.

Furthermore I want to thank my external analytic partners Dr. Lutz Wilde from Fraunhofer CNT (HT-XRD, XRR), Dr. Mandy Grobosch from IFW (XPS), Dr. M.A. Verheijen from the Dept. of Applied Physics at Eindhoven Univ. of Technology (TEM) and Dr. Manfred Schuster from Siemens CT (XRR, XFA) for their beneficial analyses, which were not available at Namlab. A special thank also goes to my friends and fellow physics students Dr. Darius Pohl from IFW (HRTEM) and Dr. Chris Elschner from IAPP (XRD, PDF) for their support in analyses and very intense discussions about important topics in my thesis. I also want to thank my colleagues Andreas Jahn, Ulli Merkel and Carola Richter from IHM for their help in preparing metal layers and the structuring of the top electrodes, when it was not possible at Namlab.

My fellow colleagues Matthias Grube, Dominik Martin, Guntrade Roll, André Heinzig and Ekatarina Yurchuk contributed with their help and after work support with funny moments. And also thanks to all the other new and old colleagues, who made the live at work extremely colorful.

Finally, I want to thank my family and friends for their never-ending non-scientific support, that allowed me finishing this thesis. Thank you all.

Bisher erschienene Bände der Schriftenreihe Research at NaMLab

Herausgeber: Thomas Mikolajick   ISSN 2191-7167

Alle erschienenen Bücher können unter der angegebenen ISBN direkt online
(http://www.logos-verlag.de) oder per Fax (030 - 42 85 10 92)
beim Logos Verlag Berlin bestellt werden.